CAMBRIDGE LIBRARY C

Books of enduring scholarly value

Technology

The focus of this series is engineering, broadly construed. It covers technological innovation from a range of periods and cultures, but centres on the technological achievements of the industrial era in the West, particularly in the nineteenth century, as understood by their contemporaries. Infrastructure is one major focus, covering the building of railways and canals, bridges and tunnels, land drainage, the laying of submarine cables, and the construction of docks and lighthouses. Other key topics include developments in industrial and manufacturing fields such as mining technology, the production of iron and steel, the use of steam power, and chemical processes such as photography and textile dyes.

Memoir of the Life of Sir Marc Isambard Brunel

Abandoning a military career, Richard Beamish (1798–1873) decided to become a civil engineer. His suitability as a biographer of Sir Marc Isambard Brunel (1769–1849) stems from the period he spent working closely with the Brunels on the Thames Tunnel. Published in 1862, this memoir recounts the elder Brunel's eventful life and work, including his youth in France, his flight to America in the aftermath of the French Revolution, his lesser-known ventures in the early nineteenth century, and the tunnelling project which would consume much of the second half of his life. An informed portrait of a figure who has since been outshone by his more famous son, this book includes first-hand accounts of the ill-fated early attempt to build the Thames Tunnel, which was abandoned in 1828 due to flooding and lack of funds, and of Brunel's vindication upon its eventual completion in 1843.

Cambridge University Press has long been a pioneer in the reissuing of out-of-print titles from its own backlist, producing digital reprints of books that are still sought after by scholars and students but could not be reprinted economically using traditional technology. The Cambridge Library Collection extends this activity to a wider range of books which are still of importance to researchers and professionals, either for the source material they contain, or as landmarks in the history of their academic discipline.

Drawing from the world-renowned collections in the Cambridge University Library and other partner libraries, and guided by the advice of experts in each subject area, Cambridge University Press is using state-of-the-art scanning machines in its own Printing House to capture the content of each book selected for inclusion. The files are processed to give a consistently clear, crisp image, and the books finished to the high quality standard for which the Press is recognised around the world. The latest print-on-demand technology ensures that the books will remain available indefinitely, and that orders for single or multiple copies can quickly be supplied.

The Cambridge Library Collection brings back to life books of enduring scholarly value (including out-of-copyright works originally issued by other publishers) across a wide range of disciplines in the humanities and social sciences and in science and technology.

Memoir of the Life of
Sir Marc Isambard Brunel

RICHARD BEAMISH

CAMBRIDGE
UNIVERSITY PRESS

CAMBRIDGE UNIVERSITY PRESS

Cambridge, New York, Melbourne, Madrid, Cape Town,
Singapore, São Paolo, Delhi, Mexico City

Published in the United States of America by Cambridge University Press, New York

www.cambridge.org
Information on this title: www.cambridge.org/9781108064965

© in this compilation Cambridge University Press 2013

This edition first published 1862
This digitally printed version 2013

ISBN 978-1-108-06496-5 Paperback

SIR MARC ISAMBARD BRUNEL

LONDON
PRINTED BY SPOTTISWOODE AND CO.
NEW-STREET SQUARE

Painted by H Wyatt

Engraved by H. Adlard

London Longman & C°

MEMOIR OF THE LIFE

OF

SIR MARC ISAMBARD BRUNEL

CIVIL ENGINEER
VICE-PRESIDENT OF THE ROYAL SOCIETY
CORRESPONDING MEMBER OF THE INSTITUTE OF FRANCE
&c. &c. &c.

BY RICHARD BEAMISH, F.R.S.

Suum Cuique

LONDON

LONGMAN, GREEN, LONGMAN, AND ROBERTS

1862

PREFACE.

IN undertaking to place on record the leading events
in the life of one of the greatest mechanists of the
age, I am fully sensible of the difficulties to be overcome, and of the delicacy with which many passages
in so eventful a life must be approached; and it is
only because no other hand has been extended to
secure some of those fleeting notices of a personal
nature, which time only too soon obliterates, that I
would now endeavour to save from oblivion the labours
of one to whom England stands largely indebted for
her progress in the mechanical arts.

True, on the one hand, our great moralist has laid
it down, that "they only who live with a man can
write his life with any genuine exactness and discri-
mination;" and from the long intimacy which I was
permitted to enjoy with Sir Isambard Brunel, I am
enabled so far to fulfil the condition required. But, on
the other hand, when I call to mind that the same
celebrated authority has also declared, "that few
people who have lived with a man know what to
remark about him," I should have been tempted to
abandon my task, had not the too flattering confidence
of the surviving relations of my esteemed friend placed
at my disposal all the documents which they have been

enabled to collect calculated to throw light upon his distinguished career. Whatever opinion may be adopted, I should not feel justified in declining to accept the duty which has been thus assigned me.

Amongst the manuscripts placed in my hands I find two short memoirs, one by the late M. de St. Amand, the other by the late Mr. Carlisle, librarian to George IV.; together with the published " Notice Historique " of M. Edouard Frère, addressed, " A l'Académie des Sciences de Rouen."

Between M. de St. Amand and the Brunel family .there had existed a long and intimate friendship, cemented not only by a common sympathy of loyalty, but by that love for mechanical and scientific pursuits for which both were distinguished. Both were in the service of their unhappy king, Louis XVI., at the period of that fearful convulsion which tore France to pieces, and shook the whole fabric of European governments; both fled from the horrors enacted in the name of liberty: Brunel to America, in the full enjoyment of health and strength, and supported by hopeful antici- pations of the future; M. de St. Amand to England, under circumstances of unusual difficulty, danger and despondency. He had received a severe wound on that memorable 5th of October, 1789, when, as one of the king's body-guard, he was hunted through the palace of Versailles. Having made his escape to the wood of Montmorenci, which could only afford tem- porary shelter, he dragged his disabled limbs to the coast, and ultimately succeeded in reaching this coun- try. Here, by a virtuous struggle against fallen cir- cumstances and an enfeebled constitution, he maintained a position amongst the good and great, alike honourable to his intellectual attainments and to his moral worth.

I may further add, that amongst many mechanical inventions of M. de St. Amand, an instrument for determining a ship's course is said to have possessed great merit, and to have deserved a better fate than room to moulder in the archives of the Admiralty. We may well understand with what deep interest M. de St. Amand's enthusiastic and sympathetic mind followed the development of Brunel's genius, and made him naturally solicitous to place upon record the successes which his friend had achieved, that they might be preserved as well from the treacherous hand of the despoiler, as from the obliterating influence of time.

For many of the anecdotes relating to Brunel's early life and social character I am, however, mainly indebted to notes made by his daughter, Lady Hawes ; to the valuable journal of the Rev. H. T. Ellacombe ; to circumstantial and detailed notices by his own pen ; and to communications with which he, from time to time, favoured me during the daily, often hourly, confidential intercourse subsisting between us during the progress of the Thames Tunnel.

CONTENTS.

—◆—

CHAPTER I.

CHAPTER II.

CHAPTER III.

CHAPTER IV.

CHAPTER V.

CHAPTER VI.

1802–1803.

CHAPTER VII.

1802–1810.

CHAPTER VIII.

1805–1816.

CHAPTER IX.

1809–1814.

CHAPTER X.

1814–1819.

CHAPTER XI.

1821–1831.

CHAPTER XII.

1814–1821.

CHAPTER XIII.

1821–1826.

CHAPTER XIV.

1824–1825.

CHAPTER XV.

1825–1827.

CHAPTER XVI.

1827–1828.

CHAPTER XVII.

1829–1831.

CHAPTER XVIII.

1831–1843.

CHAPTER XIX.   .

CHAPTER XX.

1843–1849.

CONCLUSION.

APPENDICES.

LIST OF ILLUSTRATIONS.

ERRATA.

MEMOIR

OF

SIR MARC ISAMBARD BRUNEL

CHAPTER I.

BIRTH AND YOUTH, 1769–1786.

SENT TO GISORS — STORY OF DOG — PUNISHMENT — STORY OF POR-
TRAIT—SENT TO ST. NICAISE, ROUEN — DANNECKER — TALENT FOR
DRAWING AND CONSTRUCTION — EARLY ADMIRATION OF ENGLAND
— CONSTRUCTS AN ORGAN — STUDIES FOR THE NAVY — CONSTRUCTS
A QUADRANT — ENTERS NAVY.

MARC ISAMBARD BRUNEL, the subject of this
memoir, was born at Hacqueville on the 25th of
April, 1769 ; an epoch rendered remarkable for the
birth of many celebrated men — the most conspicuous
being Humboldt and Cuvier, Napoleon Bonaparte and
our own illustrious Duke of Wellington.

Hacqueville is situated near to Gisors, in that part of
Normandy formerly called the "Vexin;" but which
has since the revolution received the appellation of
"the department of the Eure." The name of Brunel
is found at every period in the ancient records of the
province. The privilege of Maître des Postes of the

district seems to have been an inheritance of the family.
The Brunels enjoyed, however, the higher privilege of
having given to their country an unusual number of
men remarkable for their piety and learning. They
have also the honour to claim, as one of their distinguished members, the greatest painter which France has
produced. Not far from Hacqueville, at Les Andelys,
is the birth-place of Nicolas Poussin, whose mother was
of the family of Brunel. The father of Sir Isambard
was held in high esteem, not only for the simplicity
and openness of his character, but for the honourable
frugality with which he dispensed a narrow income,
and the prudence, tenderness, and diligence with which
he sought to educate a family of three children, viz.,
two sons, of whom Marc Isambard was the second, and
one daughter. M. Flahaut, in an address to the civil
engineers at Paris, speaks of the Brunels, Sir Isambard
and his son, as having sprung from the working classes :
" Sortis de la classe des artisans, ou même des ouvriers,
ils n'ont dû qu'à eux-mêmes ce qu'ils ont appris."
M. Flahaut is in error, and though few contemplations
are more gratifying, or more instructive than the successful struggles of self-taught men of humble origin,
yet we should be scarcely justified in excluding from
the catalogue of fame, those who have had the moral
courage to resist the various enervating influences which
a recognised social position only too readily produce.

Of the mother of Brunel, whose maiden name was
Lefèvre, I have been able to learn little. That her early
loss was long and severely felt, there seems to be no
doubt. An attempt was made at the earliest period to
impress Marc Isambard with the necessity of giving his
mind to the acquirement of classical knowledge, being
destined to succeed to a living in the gift of the family,

which would have secured to him a sure, though comparatively humble provision. Accordingly, he was sent soon after his mother's death, and when he had attained his eighth year, to the College of Gisors, but in vain ; literary studies possessed no charm for his tastes, and they therefore retained no hold on his affections. No efforts on the part of his teachers at school, no punishments inflicted by his father at home, could insure one half the attention to his classical studies that he spontaneously bestowed upon the carpenter's shop and the wheelwright's yard in his native village.

An event may be here related which has been recorded by Brunel, and preserved by Lady Hawes, in connection with Gisors, and which threatened to impugn his moral character at the very outset of life.

His father had accompanied him to the school, after one of his vacations, taking with him the amount of the previous quarterly payment in crown pieces. The canvas bag containing the money being emptied upon the table, was counted to the master in presence of the boy, and a receipt given in due form. A conversation subsequently ensued, the money remained on the table, and was not removed until the father had taken his leave. Before the master placed it in his strong box, however, he again counted it, when behold, some pieces were missing. Young Brunel was questioned, but he stoutly denied all knowledge of the missing money— suggesting, at the same time, that Flore might have taken them.

Now Flore was a little dog belonging to the Brunels, which had been taught many accomplishments, and which had accompanied the boy and his father to Gisors. " A dog to take money," exclaimed the master, " C'est un peu trop fort ! " Crown pieces he thought were as

little capable of being swallowed as the fanciful im-
plication of the boy, " Non, non, mon enfant, il n'est
pas possible." Still the boy persisted in denying any
knowledge of the money, and begged that his father
should be sent for, and to mind and bring Flore. At
length, with evident reluctance on the part of the
master, a letter was despatched to Hacqueville, and in
due course, father and dog made their appearance. The
cause of summons having been explained, they were
shown into the room where the money had been
counted on the previous day. Flore was now observed
to drop her tail, and to betray symptoms of embarrass-
ment. " Cherche, Flore — cherche, cherche ! " cried
papa ; but Flore would not comply. The master remained suspicious. The boy looked anxious ; the father
a little angry. At length Flore seemed to relent ; the
tail no longer drooped, — the dull eye brightened, and
she began to "*chercher*" in earnest. The master was in
amazement — the boy regained his confidence — the
father his good humour, when Flore producing the
missing coins from the corner in which she had hidden
them, solved the mystery, and at once vindicated the
integrity of the poor boy, which her accomplishment
had so nearly compromised.

The holidays were devoted to drawing and carpentry.
The old château, for centuries in the possession of the
family, and the Château Gaillard, built by Richard Cœur-
de-Lion, as a frontier defence against Philippe Auguste,
in the neighbourhood of Les Andelys, were the favourite
subjects of his pencil. This early exhibition of
artistic and mechanical talent is, perhaps, only equalled
by our own Smeaton, who, from the earliest period of
his life, gave unmistakeable evidence of mechanical
aptitude, and to which the fish in the small ponds at

Ansthorpe, his father's residence, not unfrequently fell victims ; the water being experimentally transferred from one pond to another for the gratification of the embryo engineer. For Smeaton, also, mechanical design and the construction of models had more interest than the drawing of deeds, or the engrossment of parchment, to which he had been destined by his father, but against which his nature also rebelled.

Less disposed to be guided by circumstances than the elder Smeaton had been, the father of Brunel sought to compel obedience to his wishes by the infliction of various punishments, solitary confinement being the most often employed. Of one room, selected for that purpose, the little recusant entertained something like horror. On the walls of that room hung a series of family portraits. Amongst them was one of a grim old gentleman, the eyes of which appeared to be always turned towards him, with a frown so stern, menacing and forbidding, that fear and vexation took possession of his mind. No matter in what part of the room he took shelter, still those angry eyes were upon him ; nor could he resist their painful attraction, for look at them he must. His nervous temperament, becoming unable to bear the sort of persecution any longer, he one day, when nearly distracted, collected all his strength to drag a table from one end of the room, and to place it immediately beneath the picture. Upon the table he contrived to lift a chair, and on this chair he climbed. Regardless of consequences, he at once revenged himself for the misery he had endured by fairly cutting out the eyes from the canvas with the aid of his friendly pocket-knife.

The early life of Dannecker, the celebrated sculptor, offers a similar example of the fruitless attempt to check, if not destroy, the early impulses of genius. With Dannecker it was not the wheelwright's, but the stonemason's yard that proved the attraction. This yard was contiguous to his father's house, and there he found abundant facility for the gratification of his taste; and there it was certain he would be found chipping stones when missed from home, or from the stables of the Duke of Würtemberg, where he was employed under his father. Punishment followed punishment to no purpose. Solitary confinement was resorted to with as little success as it had been with Brunel. It is related that, one morning, having made his escape through the window of his prison, he presented himself, with four companions, before the duke, to whom they preferred a petition, praying that they might be permitted to enter a school, which the duke had recently established, for the benefit of the children of his servants, and in which drawing and music, as well as the ordinary elements of knowledge, were taught free of cost. To Dannecker's inexpressible joy his prayer was heard, and he was at once relieved from parental tyranny and ignorance; which, although powerless to destroy the instinct of the boy, would have been productive of years of pain and sorrow to his ardent and sensitive nature.

Less fortunate than Dannecker, Brunel continued to be subjected to every repressing influence. At eleven years of age he was sent to the seminary of St. Nicaise, at Rouen—one of the numerous establishments connected with the large ecclesiastical college of that town,—still with the hope of securing him for the church. But nature was not to be turned from her course. So strongly did his taste for drawing continue to exhibit itself, that the superior was unwilling to deprive him of the advantage of a master. His first lessons were directed to the delineation of the human figure,—and more particularly of the human countenance. But, unable to endure the tedium of repetition, and the routine of copying each feature in detail, he produced one morning—no less to the astonishment of his master than the admiration of the superior—a finished portrait of an individual well-known, and in which the distinguishing traits were said to be admirably expressed.

In this juvenile effort will be recognized the leading characteristics of Brunel's mind,—largeness, and comprehensiveness of conception, combined with the utmost accuracy of detail. The several features, upon which routine would have pondered for weeks, as distinct and isolated facts, were at once combined, and made to subserve the general purposes of a portrait, which the young artist presented to his master, as the best vindication he could offer for declining any further instruction at his hands.

His sources of amusement differed widely from those of other children of the same age. At a period when most children can scarcely manage an ordinary knife, young Brunel was familiar with the use of the greater part of the tools found in a carpenter's shop ; and so inveterate was his love of such tools, that he has been known to pawn his hat that he might possess one newly exhibited in a cutler's window at Rouen, though at the time unacquainted with its special use. At the age of twelve he constructed various articles with as much precision and elegance as a regularly educated workman. Every day showed the rapidity with which Brunel could seize upon all combinations of material forms, and exhibited some new feature in his aptitude for mechanical pursuits. The construction, the rigging, and the motive power of vessels early attracted his attention ; and the drawings which he executed at this period are said to have been models of form and detail.

Of all the mechanical operations which he witnessed in those early days, the one which excited the largest amount of interest, was the manner in which the tire of a wheel was fixed to its rim or felloe ; indeed, a carriage wheel seemed, to the latest period of his life, to excite in Brunel renewed delight. Its simplicity. beauty, and perfect mechanical adaptation, always called forth his unqualified admiration.

About this period, in one of his daily visits to the quay at Rouen, a locality which had for him great attractions, his attention was more than usually excited by two cast-iron cylinders which had been just landed, and which, when compared with his own height, for he always formed a mental scale, seemed to him gigantic. What their use ? Whence they came ? Whither they were going ? were questions to which for some time he in vain sought replies. At length. a boatman alongside the quay, interested in the lad's eagerness, beckoned him to descend, and promised to afford the wished for explanation.

It may well be conceived with what alacrity the invitation was accepted, and how joyously that boat was entered. With what earnestness Brunel listened to the explanations of the friendly boatman. How those cylinders were part of a fire engine (then so called) for the purpose of raising water ; that they had just arrived from England, where many such things were made. " Oh ! " exclaimed Brunel, " quand je serai grand, j'irai voir ce pays-là."

On another occasion the superiority of the work-

manship of the different parts of a carriage recently
landed on the quay attracted his observation. "Ah!"
he exclaimed, "qu'ils sont habiles dans ce pays-là ; j'irai
le voir quand je serai grand."

With a mind so much alive to everything into
which construction entered, it was no wonder that
Brunel's imagination should have been aroused by the
mechanical arrangements of musical instruments.
Having taught himself the flute, and the construction
of the harpsichord, the possibility occurred to him of
combining the effects of both in one mechanical
arrangement, and this, without any knowledge of the
laws of sound, or the rules of art. He thus, uncon-
sciously, rivalled the ingenious inventions of Vaucan-
son, of whose name and success he was equally
ignorant, and of the self-taught peasant who erected,
at Moshuus, in Norway, an organ, described by Sir A.
de Capell Brook as "perfect in its parts, and with a
variety of stops."

Our own Watt, in the early part of his career, and
without any knowledge of the science of music, or
correct appreciation of musical intervals, turned his
mechanical skill in a similar direction. "He constructed guitars, flutes and violins, and proposed a
mode of playing on the musical glasses which should
be independent of the wetted finger. In organ building, also, Watt introduced many valuable improvements, such as delicate indicators and regulators of
the strength of the blast ; and ultimately he was enabled to establish the theory of Daniel Bernouilli
relative to the mechanism of the vibration of musical
chords, and which explains the harmonious sounds
that accompany all full musical notes." *

* Life of Watt, by J. P. Muirhead, M.A.

Brunel, though he did not aspire to the construction of an organ, nor to the attainment of a knowledge of the theory of music, nor the principles of harmony, yet accomplished what was, perhaps, a more wonderful feat, considering his age (eleven years), than even that which had been performed by Watt in his twenty-third year, after long experience as a professed mechanist. Unfortunately, this interesting model of Brunel's musical machine no longer exists, and therefore we have no means of determining how far it embraced the requirements of the beautiful and ingenious instrument known as the barrel organ. But these exhibitions of mechanical precocity afforded little consolation to a parent whose mind was occupied with the grateful anticipation of seeing the family living still occupied by a Brunel. " Ah ! mon cher Isambard," he used to say, " si tu prends ce parti-là, tu végéteras toute ta vie." It must, however, be remembered, that at the period of which we speak, there was nothing to suggest the changes about to take place in the industrial arts ; nothing to indicate that rapid development which the application of steam. as a motive power, was destined to produce. In Rouen there did not exist one cotton spinning-machine. The only one to be found in the country was at Louviers, although indeed it is recorded by Dr. Royle, in his " Productive Resources of India," that the Rev. W. Lee, of St. John's College, Cambridge. the inventor of a machine for knitting and weaving stockings, was induced by Henry IV., just 200 years before, to establish himself at Rouen, because he received no encouragement at home. The populace of Rouen had, however, now, in their ignorance and blindness, opposed every attempt to introduce spinning-machines, or to erect manufactories for muslin or muslinette. So late as 1787 cotton was spun by the hand in Rouen, and throughout the province. In 1789 some speculative persons ventured to import machinery from England, but it was quickly demolished by the artisans.*

No wonder that a father, entirely ignorant of the value of mechanical appliances, which were then, and long continued to be, unappreciated either by society or by government, but who was perfectly alive to the secular as well as spiritual power of the clergy, should witness, with profound disappointment, the growing tendencies of his child.

The efforts of his father, aided by zealous and accomplished teachers, having failed to wean the young artist from his mechanical pursuits, he solicited, and at length obtained permission to visit an old friend of the family at Rouen,— M. François Carpentier,— under whose direction he systematically studied drawing and perspective. To these studies hydrography was added, with a view of qualifying him to enter the navy, a service for which he had

---

* The rude self-protection which urged the natives of Rouen to raise their hands against machinery that they believed was destined to rob them of their bread, can be better understood, and more readily justified, than the intolerance and learned bigotry of those claiming the highest social position and authority, in enlightened Scotland, in the early part of the eighteenth century. Amongst other examples of the profound ignorance of art, and the fierce religious fanaticism which characterised that period, Mr. Robert Chambers (*Domestic Annals of Scotland*) notices the manner in which the inventor of the first agricultural machine was received in 1737. It was denounced as the " new-fangled machine for dighting the corn frae the chaff; thus impiously thwarting the will o' Divine Providence by raising wind by human art instead of soliciting it by prayer, or patiently waiting for whatever dispensation of wind Providence was pleased to send upon the shieling hill."

exhibited a strong predilection. The relief which the new course of study afforded him was often alluded to in after life. Under M. Dulagne of Rouen, the learned author of a treatise on hydrography, which forms a supplement to those of MM. Bouguer and De la Caille, the propositions of Euclid had only to be stated to be understood : demonstration was neither asked for nor required. After the third lesson in trigonometry he proposed to his astonished and delighted master to determine the height of the spire of the cathedral. " Il l'admit," says Brunel in a letter to a friend, " je fis de suite un instrument, assez grossier à la vérité, mais assez juste, pour confirmer la théorie et la pratique."

His love for construction still continued to afford the highest gratification to his leisure hours, and the models of vessels which he produced are said to have possessed singular beauty of form and finish.

His industry, his intelligence, the integrity of his mind, and the sweetness and loyalty of his disposition, endeared him to all with whom he became associated. So conscious had M. Dulagne become of his pupil's superiority, that he joyfully seized the opportunity to procure for him the notice of the Minister of Marine, the amiable Maréchal de Castries, upon the occasion of his visit to Rouen, in the suite of Louis XVI., when on his return from Cherbourg, and upon whom Brunel made so favourable an impression, that the marshal was induced to nominate him " Volontaire d'honneur," before the usual time, to the corvette " Le Maréchal de Castries." However painful the feeling of disappointment to his father may have been at the failure of his favourite project to secure so much talent to the church, that regret must have been greatly modified in receiving from M. de Castries an assurance of protection for his child, and in knowing that the honour conferred upon him had only once before been granted, and that to M. de Bougainville, the celebrated circumnavigator.

As an illustration of the accuracy of the observing and constructive powers of Brunel at this early period, it may be here further stated that, when introduced to the captain of the vessel in which he was to sail, an instrument on the table attracted his attention. This was a Hadley's quadrant. He had never seen one before, and was now simply told its use. He did not touch it, but, walking round the table, carefully examined it. In a few days he produced an instrument of his own construction. "Assez grossier, à la vérité," as he used to say ; "mais assez juste ;" his only theoretical guide being a description of the instrument, in a work on navigation, supplied to him by his master. But this first attempt only stimulated the young mechanist to further efforts, and, with the unwilling aid of a few crowns from his father, he executed another quadrant in ebony with so much accuracy that, during the whole period of his connection with the navy, he required no other. When it is remembered that this instrument demands in the constructor a knowledge not only of geometry, trigonometry and mechanics, but of optics, one is filled with astonishment and admiration at the intuitive sagacity which brought all this knowledge to bear upon so delicate and complicated a construction. When about to embark in the new career which his conduct and his talents had opened to him, Brunel was attacked with small-pox. Some months' delay seems however to have been the only drawback. Upon his recovery he joined his vessel, destined for the West Indies, it may be presumed under the same favourable auspices.

CHAPTER II.

1786–1793.

PERIL AT PARIS — MISS KINGDOM — DISTURBANCES AT ROUEN — QUITS FRANCE — FORGES PASSPORT — INSURRECTION IN ST. DOMINGO — LANDS AT NEW YORK — CONNECTION WITH M. PHAROUX.

THE marine of France had attained to an unprece-dented pitch of efficiency and power under the fostering care of Louis XVI. Her flourishing colonies in the Antilles still afforded a valuable nursery for her seamen. For although, at the close of the war of in-dependence (1763), there only remained to her, of all her great possessions in the west, the Island of St. Domingo, and a few of the lesser islands, yet in value they equalled, if they did not exceed, the colonial posses-sions of all other nations taken together.*

---

* From St. Domingo alone the exports amounted to

	168,000,000 francs =	£6,720,000
And the imports to	250,000,000 francs =	£10,000,000
		£16,720,000

While at the same period, the commerce of Great Britain with her American and West Indian colonies did not exceed £8,288,145.

	Exports.	Imports.
British America	1,119,991	255,797
West Indies	2,784,310	4,128,047
	3,904,301	4,383,844
Total		£8,288,145

From 1786 to 1792 Brunel seems to have been actively engaged in his profession ; and from his intelligence, gaiety, amiability, and general refinement, to have endeared himself, as well to his superior officers as to his ruder companions. It may be said, indeed, that he was a universal favourite ; a *jeu de mot* upon his names offers some evidence of the light in which he was regarded. On board he always went by the title of *Marquis* (Marc I—sambard).

It is much to be regretted that there remains no record of the impressions which the susceptible mind of Brunel received of those countries and peoples with which, during the six years of his naval service, he must have been brought into contact. In January, 1793, we find him in Paris, his ship being paid off. There, events were succeeding each other with a rapidity and violence unparalleled, perhaps, in the annals of human passion ; and, on the very day when the Convention pronounced sentence against the unfortunate Louis XVI., Brunel was found defending his own loyal opinions in the colonnade of the *Café de l'Échelle*, little conscious of the risk to which he subjected himself. In the heat of discussion, and in reply to some ferocious observations of an ultra-republican, he more boldly than prudently exclaimed, " Vous aurez bientôt à invoquer la protection de la Ste Vierge, comme autrefois ' à furore Normannorum libera nos Domine.' " *

Fortunately for our young loyalist, M. Taillefer, a member of the Assembly, by committing an act of greater indiscretion, turned the attention of those pre-

* Inscription on one of the Gates of Rouen, after the city had been taken by the French.

sent upon him, and, in the confusion which ensued, Brunel was enabled to effect his escape. That night he slept at the *Petit Gaillard-bois* next door, and the following morning at an early hour quitted Paris.

At Rouen, where his family had been known to entertain moderate views, Brunel was enabled to remain for a time undisturbed : but at a period when every species of despotism was exercised without a despot being acknowledged ; and when, "to stifle every emotion of sensibility," was, according to Robespierre, the greatest proof which a man could give of devotion to his country, it was not possible that France could any longer offer Brunel a home.   And though the death of Louis XVI., which took place four days after Brunel's escape from Paris (January 21st, 1793), was quickly followed by an indiscriminate and mutual massacre of the judges and executioners of that ill-fated prince, yet was there no safety, either for the loyalist or constitutionalist, under the then existing jealous and unprincipled government.

At Rouen, Brunel again availed himself of the protection of his relative, M. Carpentier, and it was when under his hospitable roof that an event occurred which will be found to have exercised a marked influence upon Brunel's future career.   In that house, for the first time, he met a young English lady, of the name of Kingdom, gifted with no ordinary personal attractions.   This lady was the youngest of sixteen children, nine of whom only reached maturity.   Her father, who had been an army and navy agent at Plymouth, was dead, and her widowed mother, supported by the active interest of the member for Plymouth, who had been left guardian to her children, was enabled to secure provision for her sons in the navy office.   Solicitous to obtain every advantage for her favourite daughter Sophia, who had just attained her sixteenth year, she was induced to accept an invitation from some West India friends, M. and Madame de Longuemar, to permit the young lady to accompany them to Rouen, that she might acquire a practical knowledge of the French language. It might appear to be matter of some surprise that Miss Kingdom should have been permitted by her friends to enter France at all at a period (1792) when everything was tending so rapidly to a political crisis, if we were not aware how little was generally known in England as to the condition of political parties in France. But already royalty was in captivity, and the most fearful cruelties were being committed in the name of liberty. Circulars had been addressed by the municipality of Paris to the other cities of France, inviting them to imitate the massacres of the capital. At Rouen two young ladies, known to M. and Madame de Longuemar, were dragged into the street by the insensate mob, and with shouts of " à la lanterne " were actually murdered, because they had been heard to play a loyalist air on their pianoforte. The alarm thus created in Rouen hastened the departure of M. and Madame de Longuemar for the West Indies. Miss Kingdom would gladly have accompanied them, had not a severe illness rendered her unable to encounter the inconveniences of a sea voyage, and she was in consequence left under the care of M. Carpentier, the American consul, the intimate friend of the Longuemars, himself married to an English lady, and, as we have seen, the relative and tried friend of Brunel. Here it was, then, that Brunel became acquainted with Miss Kingdom. Opportunities were not wanting for the cultivation of an acquaintance in which mutual sympathy awakened mutual admiration. For

c

Brunel beauty of form possessed an irresistible attraction. One day, while the young lady was admiring his first attempt at oil painting—still in existence,—and, in her graceful and winning manner, pointing out what parts pleased her most, he turned to Madame Carpentier, and whispered, "Ah! ma cousine, quelle belle main!" "Oui," she replied, "mais elle n'est pas pour toi." Not long after this little event, an *émeute* of the republican party called out the royalists to suppress it : Brunel amongst the number. The excitement was tremendous—the danger great. It was no wonder, therefore, that love should take the place of admiration and sympathy; and that a reciprocal avowal of passion should be the consequence. When all the houses of the respectable inhabitants had to be barricaded against the intrusions of the *sans-culottes* or *bonnet rouges*,—when the distant roll of the drum brought its mysterious forebodings of some violent display of popular paroxysm,—or the clang of the tocsin summoned the loyal and well-disposed for the protection of property and life,—when surrounded, in short, by perils the most appalling, the thoughts of these two loving hearts would necessarily be concentrated each upon the other, and impressions would be received which neither time nor circumstances could ever efface. Young Brunel's position now became daily more critical; a longer delay in Rouen might be dangerous. Already a new phase in the revolutionary development had presented itself. Provisions and public money, destined for the army, had been intercepted, and everything portended another fearful catastrophe. The Jacobins had prevailed—the reign of terror had commenced—the Convention was prostrated—its power had passed to the committee of public safety!—to Robespierre, St. Just, Couthon,

Collot, and "to the ignoble, sanguinary, and depraved Barère."

A column of Federalists had issued from Britany and Normandy, with the view of marching upon Paris, while other columns from Bordeaux, and the basin of the Loire,—from Avignon and Languedoc, Grenoble, the Ain, and the Jura, were pressing forward towards the same point, with the avowed object of rescuing the republic from the sanguinary tyranny of its own children.

Upon the plea that he was engaged to purchase corn and flour for the army, Brunel with difficulty obtained a passport to America, its operation being limited to one year.

No time was to be lost. On the 7th July, 1793, he bade adieu to his native France; not, as we may well believe, without feelings of deep and heartfelt sorrow. But his loyal spirit could never have allied itself with those whose hands were imbrued in the blood of their sovereign.

All hope of entering upon any other career than that of war was too far distant to afford any reasonable prospect of employment in his native country, and thus did the iniquity of her government deprive France of the services of one of her most gifted sons.

The attachment, also, which Brunel had formed, while it tended still farther to embitter his farewell, must yet be regarded as adding another motive for entering upon that struggle for independence upon which he had resolved, and as offering a new and powerful stimulant to the exercise of faculties which he must have felt conscious of possessing, in the hope of ultimately winning a prize upon which his imagination and his affections had set the highest value.

That Miss Kingdom was strongly impressed with the devotion which she had inspired, her constancy and fortitude, through many years of trial, afford the most unequivocal testimony.

At Havre, Brunel found an American vessel called "Liberty," about to sail for the United States, in which he secured a passage. Scarcely had he congratulated himself upon his providential escape from tyranny and oppression, when he discovered that the passport, to obtain which he had devoted many anxious hours, had been forgotten. The first pang of disappointment passed, no time was given to vain regret; a mind so full of resource as was that of Brunel, could scarcely fail to find some means by which the loss might be supplied; a loss which, to any other, would have proved absolutely irreparable, and might have proved fatal.

Having obtained from one of his fellow passengers the loan of his important credential, he very soon produced a copy, so admirably executed in every minute detail, even to the seal, that it was deemed proof against all scrutiny.

To his caligraphic skill was he now indebted for freedom, and perhaps for life. Scarcely was the ink dry, when a French frigate hove in sight. A signal was soon after made for all the passengers on board the American vessel to parade on deck, that their passports might be examined. Any detected irregularity would have subjected Brunel to the humiliation of arrest, and his immediate transmission back to France as *suspect*. Confiding in his artistic skill, and feeling the importance of suppressing all appearance of hesitation or misgiving, he was the first to present his bold but well simulated document, and to receive the necessary confirmation of its legality, not the slightest suspicion having been aroused as to its authenticity.

Without farther let or hindrance, he landed in safety at New York on the 6th September, 1793. There, to his dismay, he found the French squadron which had conveyed all those who were so fortunate as to escape the fearful massacre at St. Domingo.

It will be remembered that, in 1790, the Constituent Assembly had empowered each colony belonging to the republic to make known its wants, on the subject of a constitution, through its own assembly, to be elected by its own citizens. The mulatto population of St. Domingo naturally claimed to participate, as citizens, in the privilege thus heedlessly decreed, and as naturally was their claim resisted by the whites, who, as the proprietors of the greater portion of the property of the island, and the inheritors of the wealth, the luxury and the prejudices of their fathers, felt their dignity compromised and their power endangered by this insensate delusion of the Assembly.

Though inferior in point of wealth, the mulattos were far superior in point of numbers, and under the name of *Petit Blancs*, were rising into social importance ; they therefore rejoiced in the opportunity now afforded them to secure political position also.

Absorbed by class and personal contentions and animosities, the condition of the slave population was entirely overlooked.* The effect of the energetic and

* The relative proportion of the population given by Mackenzie as quoted by Alison, was, whites 40,000, mulattos 60,000, blacks 500,000. Annual Register gives, whites 42,000, mulattos 44,000, slaves 600,000.

The ruinous effect of the mistaken legislation of the Constituent

active Jacobin missionaries upon the ardent and ignorant minds of the negroes was not appreciated. More
circumspect than had been their countrymen of Jamaica,
when, in 1760, they sought to cast off the British yoke,
the negroes of St. Domingo, under their able chiefs
Brasson, Toussaint, and Hyacinthe, successfully accomplished their project, and in June, 1793, after a series
of atrocious cruelties, Cape Town, the last stronghold
of the planters, was reduced to ashes, at the time when
the whites and mulattos were actually engaged in
civil contention upon a question of privilege and caste.
" Thus fell the Queen of the Antilles," says Alison
(*History of Europe*), " the most stately monument of
European opulence that had yet arisen in the New
World; and thus democratic France, by an improvi-
dent and reckless encouragement of freedom, lost her
most valuable West Indian possessions, as constitutional
England lost her American colonies by an equally
wilful, intolerant, and perverse legislation."

The greater part of the fugitives from that devoted
Assembly will be at once seen in the following comparative commercial statistics of St. Domingo.

	A.D. 1789.		A.D. 1832.
Population . . .	686,000	. . .	280,000
Sugar exported :			
White 47,516,531 ⎱ Brown 93,573,300 ⎰	141,089,831 lbs.	. . .	none
Coffee . .	76,835,219 lbs.	. . .	32,000,000
Cotton . .	7,004,274 lbs.	. . .	none
Ships employed in Trade	1680	. . .	1
Seamen . . .	26,770	. . .	
Indigo . . .	758,628 lbs.	. . .	

Besides many other articles, such as hides, molasses, and spirits, to
the amount of 171,544,666 livres.—*Annual Register.*

Exports to Fránce .	£6,720,000	. . .	none
Imports from do. .	£9,890,000	. . .	none

Mackenzie's St. Domingo.

country had sought shelter in the United States, and however well disposed our young *emigré* may have felt to sympathise with their fallen fortune, his own personal safety called for all his attention and care.

The crews of the French vessels Brunel used to describe as so many sets of banditti. Many of them came to witness the landing of their exiled countrymen, and with coarse jests and imprecations, threatened to hang them all as a cargo of royalists; and as Brunel was personally known to many of the officers of the squadron, there was the additional apprehension that he might be recognised, treated as a deserter, and compelled, perhaps, to return to the country from which he had with so much difficulty succeeded in effecting his escape.

Having found temporary protection in the lodging-house of " one Wilson," in Hanover Place, New York, Brunel lost no time in making his arrangements for quitting the city.

A stranger in the land, he knew not where to direct his steps. To free himself from his present difficulties was his great consideration. In his dilemma he fortunately called to mind that two of his *compagnons de voyage*, M. Pharoux and M. Desjardins had proceeded to Albany, for the purpose of organising, on the part of a French company, the survey of a large tract of land near Lake Ontario, extending between the 44th parallel of latitude, and the course of the Black River; and comprehending upwards of 220,000 acres. Brunel resolved to seek them, in a vague hope that he might be permitted to bear a part in an expedition which promised abundant exercise for his enterprising spirit, and an ample field for the development of his genius: whilst it held out some prospect, if not of immediate remuneration, yet of the means of permitting him to husband his limited pecuniary resources for future emergencies.

M. Pharoux, the director of the expedition, an architect and surveyor of considerable repute, received our adventurous *emigré* with all the courtesy of a kind and generous nature. During the voyage he had already been favourably impressed with the originality of thought, and amiability of character which distinguished Brunel, and he at once felt the importance of securing the co-operation of one to whom difficulties and dangers promised to be only incentives to exertion, and the means of drawing forth natural resources of no ordinary kind.

Accompanied by four Indians, supplied with two tents, a few axes, and fowling-pieces, these three enterprising French gentlemen entered upon the arduous duty, not only of exploring, but of actually mapping a region hitherto scarcely known—a region where nature had for ages put forth unrestrained her power and her beauty—ere it had been brought within the bounds of civilisation. The glories of the physical world were appreciated by Brunel in their widest extent, and the impressions made by the richness, variety and magnitude of the vegetation of those primeval forests was ever remembered by him with renewed pleasure, mingled with a certain awe, when he called to mind the perils and the gloom by which his path had been so often compassed. Leaving our young adventurer to pursue his novel avocations, and to develope his newly discovered faculties, we shall return to Rouen, where the reign of terror was holding its court.

CHAPTER III.

1793–1799.

MISS KINGDOM IMPRISONED — TRAVELLING IN AMERICA, 1793 — FRENCH EMIGRANT FAMILY — OJIBBEWAY CHIEF — MR. THURMAN — ENGINEERING TALENT — PLANS FOR SENATE HOUSE, WASHINGTON — PARK THEATRE, NEW YORK — LOCOMOTIVE WINDMILL — CANNON FOUNDRY — DEATH OF M. PHAROUX — ADMITTED CITIZEN OF NEW YORK — DECLINES TO RETURN TO FRANCE — NAVAL SUCCESSES OF ENGLAND — BLOCK MACHINERY FIRST SUGGESTED.

NO sooner had England entered into the coalition with the continental powers against France, than all communication between France and England was at once cut off, and the English then found on French soil were, without regard to sex or age, hurried away to prison. At Rouen, the house of the American consul was found to be no protection.

Fortunately for Miss Kingdom, the prisons were already full to overflowing ; she was, therefore, with some others, conveyed to a convent, and placed under the surveillance of the nuns. The fare was wretched, and the lodging miserable. Black bread of the coarsest kind, with pieces of straw mixed with the dough, constituted the principal food ; while the beds were formed of boards, with a billet of wood for a pillow. Still, the sympathy and kindness exhibited by the poor nuns, and the relief which she experienced in having companions of her own sex, offered some compensation to Miss Kingdom for so much physical discomfort and privation. The little luxuries, also, which the friends of the nuns would, from time to time, convey within the walls, were appreciated at no ordinary value. The little cream jug, so often filled by the trusty old servant of the Macnamara family in the neighbourhood, whenever opportunity offered, is still retained as a memorial of the sufferings and the sympathies of that iniquitous period.

Much of the time not devoted to religious observances was employed by the nuns in the cultivation of the arts ; and Miss Kingdom was indebted to the instructions she obtained in this convent for her dexterity in the manufacture of artificial flowers.

Many times, during her confinement, had Miss Kingdom to witness the loss of some of her companions, condemned to the guillotine, not knowing when her own turn might arrive. At length, the hope of rescue died out ; death was casting his dark shadow before him, and despair had taken possession of the hearts of the small remnant of her companions, when behold, one morning in July 1794, the doors of the convent were thrown open, and they were declared free to depart whither they would.

Stunned by so unlooked for a reprieve, they were utterly unable to realise the fact that the arch tyrant of the revolution no longer lived, and that the reign of terror had ceased. The joy of M. and Madame Carpentier was unbounded. With open arms they received their young friend, and, as the best service they could now render, they lost no time in obtaining for her a passport to her own country.

Brunel, happily unconscious of what was passing in France, continued to devote himself to the duties of his own profession, stimulated and supported by the hope of one day placing himself in a condition to claim the object of his affections, for whose sake he desired to consecrate

" In worthy deeds each moment that is told."

Unfortunately, I have been unable to obtain any notes or correspondence relative to the eventful *coup d'essai* of his engineering life.

Communications with Europe were difficult, tedious, and expensive. I have, however, often heard Brunel speak of his sojourn in America as a period of pleasurable excitement, enhanced, perhaps, by the dangers as well as the difficulties overcome.

The only channels of communication which at that time existed between New York and its northern and eastern frontier, were by Lakes Champlain and George ; and by the Mohawk and Wood's Creek Rivers, the Oneida Lake, and the Onandago River, to Fort Oswego, on Lake Ontario.

At Albany, a hundred and forty-five miles from New York, the difficulties commenced. A waggon road for sixteen miles brought the traveller to Shenectady. From thence up the Mohawk River to the little falls, a distance of sixty-five miles, was performed in *bateaux*— light flat-bottomed boats, pointed at the ends, weighing about fifteen hundred-weight each, and worked by two men with paddles and setting poles. At the little falls occurred the first portage, or land-carriage, which led over a marsh for about a mile. To accomplish this, the *bateau* was landed and placed on a sort of sledge—an adaptation by a German colony—and so drawn beyond the falls, where the water-carriage was again resumed for about fifty miles, when another portage of six to eight miles, dependent upon the season, occurred. This brought the traveller to the Wood's Creek River, where the labours of transport were remitted. For a distance of forty miles, this beautiful river pursues its gentle course to the Lake Oneida ; from the eastern end of which the turbulent Onandago breaks its way, for about thirty miles, over rapids and rocky falls to Fort Oswego, on the Lake Ontario. This fort was one of a chain of forts which extended from the source of the St. Lawrence to the Mississippi, by which the French had, at one period, sought to deprive the English colonists of half their possessions. We have the testimony also of two English travellers, as to the condition of the public thoroughfares about this period, south and north of New York.

Mr. Francis Baily, one of the distinguished founders, if not the originator of the Royal Astronomical Society, describes in his journal, 1796-7, a journey from Baltimore to Philadelphia, in company with Mr. Ellicot, the government surveyor of the United States, whose influence appears to have first directed Mr. Baily's attention from the sublunary interests of the Stock Exchange to the glorious contemplation of astronomical philosophy.

The public conveyances seem to have been very similar in character to the *char-à-banc* of the present day. " An open coach on springs. with leather curtains, fitted up with four seats placed one before the other, suspended from the top, capable of being raised or lowered, and each seat capable of accommodating four persons ; so that the whole of the passengers face the horses." The roads were almost impassable.

" We did manage," says Mr. Baily, " to get twelve miles before breakfast : about thirteen miles between

breakfast and dinner; and about twelve more miles
before supper; having walked nearly half the way, up
to our ankles in mud." Occasionally the coach was
fairly " bogged," and left for the night.

Mr. Isaac Weld also, who travelled through the
States in 1795-6-7, describes his journey from Albany
to Lake Champlain, called by the Indians' *Caniad-eri
Quarante*, mouth, or door of the country.

The carriage, which after much difficult negotiation
Mr. Weld and his companion were enabled to obtain,
and which the proprietor boasted " was the very best
in Albany," had no springs, and was little better than a
common waggon. The traces frequently broke, and the
bridles as frequently slipped off the horses' heads. In
traversing one causeway, near Fort Edward, the over-
taxed animals were unable, without assistance, to extricate the wheels of the vehicle from between the
partially decayed trees, of which the road was formed.

From Albany to Skenesborough, a distance of *forty
miles*, occupied *twelve hours*, and the last *twelve miles*
no less than *five hours*. Well may Mr. Weld speak of
the contemplated connection of Lake Champlain with
the north, or Hudson River, by the improvement of
the navigation of Wood's-Creek River, already suggested
by MM. Pharoux, Desjardins and Brunel, as the most
important project of the day.

If, then, the ordinary route presented difficulties,
those to be overcome in the progress of exploration
may be partially conceived. By indefatigable perseverance, and the display of no common resources on
the part of their young assistant, the object of MM.
Pharoux and Desjardins was accomplished.

During this exploration, a little incident occurred
which made a lively impression upon the mind of

Brunel, and to which he never afterwards alluded but
with emotion.

As the party were rounding a small creek on the
Black River, in their canoe, where the rich luxuriant
foliage kissed the surface of the water, and entirely shut
out the banks from view, children's voices were distinctly audible, "Viens papa — viens mamma, voilà
un bateau." Who shall describe the effect of these
simple sounds upon the hearts of the exiled travellers,
as they broke the silence of an American solitude?
What visions of home-recollections must have been
presented to their affections, and with what eager
interest they must have sought the dwelling of their
expatriated countrymen, may be conceived but cannot
be described.

There, in the back woods, was a family that had
fled, as they had, from the horrors of the revolution,
supporting themselves by the work of their own
hands, and indebted to the forbearance and kindly
natural instincts of wild and lawless Indians, for that
life and that peace which had been denied to them at
home.

The same confidence with which this family had
been treated by the Indians, was extended to Brunel
and his companions. The friendly character of this
intercourse was curiously illustrated, so lately as 1845,
upon the visit of some of those people to England;
when Lady Hawes (Sir Isambard's eldest daughter)
took the opportunity of inquiring of the young Ojibbeway chief Ka-ge-ga-go-boo, whether he had ever
heard of a white man called Brunel, who visited his
country long ago. "No;" he replied, "but I have
heard my grandfather talk, with pleasure, of a wonderful white man called Bru-né." As these people always

drop the final consonant, the name would appear to be identical.

Returning to Albany, the party took their passage on board a sloop for New York. The vessel was run upon a sand-bank, and detained two tides. When about to resume her voyage, "un homme sage," as Brunel described Mr. Thurman, an American loyalist and a merchant of New York, came on board.

This gentleman had always exhibited a strong sympathy for the loyalists of France; often solacing them in their sorrows, and ministering to their wants. With M. Pharoux and young Brunel he readily fraternized; and ere the voyage to New York had been accomplished, he had engaged them to survey a line for a canal to connect the River Hudson with lake Champlain.

The engagement with Mr. Thurman became, therefore, the turning-point of Brunel's life. He had intended to return to his own country, should tranquillity be restored, and a constitutional government be established; but the fortuitous connection with this "homme sage," determined his destiny. France, and her brilliant naval service, was abandoned for America, and the humble profession of a civil engineer.

The name of Thurman is still remembered with reverence in New York, as that of one who, by promoting internal communications, tended best to develope the resources of his country.

To M. Pharoux was confided the conduct of the operations; but as difficulties accumulated, the superiority of Brunel's genius became so apparent, that with a mind as liberal as enlightened, M. Pharoux did not hesitate to resign the command into the hands of his more gifted companion; and thus was Brunel, by the force of his character, and the influence of circum-

stances,— mysterious powers, which, under the name
of accidents, are so often found to direct the destiny of
human life,— placed in the position best calculated to
promote his own happiness, and to confer lasting
benefit upon his kind.

Brunel's attention was now directed, not only to the
projection of canals, but the improvement of the navigation of rivers.   His ingenuity soon suggested the
means of freeing the beds from masses of rock and
embedded trees ; and, by lateral cuts, of evading falls
and cataracts, which rendered navigation not only dangerous but often impracticable.   He may therefore
be considered as the pioneer of those great inland
communications, which have tended so largely to promote the commercial prosperity of the States.

The connection so auspiciously formed with Mr.
Thurman, opened to Brunel other and more brilliant
opportunities for the exhibition of his constructive
talents.   Success attended all his efforts ; and thus, in
the course of less than twelve months, he had achieved
a name and secured an independence.

The building which served as the great council-
chamber of the nation at Washington possessed neither
the accommodation which the increasing business of
the States required, nor the architectural dignity which
the majesty of Congress demanded.   It was therefore
resolved that architects should be invited to send in
plans for a new structure, which would be subjected
to competition.   Amongst the competitors appeared
Brunel and his friend M. Pharoux, an architect, it
must be remembered, by profession ; but so superior
in arrangement, elegance, and grandeur of design
were the plans of Brunel, that the judges were relieved
from all difficulty of selection.   Principles of economy,

however, interfered ; and while they robbed the nation of a noble structure worthy of its greatness, they also deprived Brunel of that honour and those emoluments to which his attainments and his skill entitled him. Fortunately the time and talents which he had displayed in this new field of art were not suffered to be lost. Plans were soon after demanded for a theatre in New York. With considerable modifications of the former design, Brunel's were accepted.

M. Pharoux again entered into competition ; but so far from the success of his friend exciting the slightest jealousy or ill-will, he was amongst the first to offer his congratulations, and to solicit as a favour that some of the decorative portion of the work might be accorded to him ; not only that the friendship which circumstances had so happily established might he perpetuated, but that he might also secure the privilege of " free admission." Brunel and Pharoux were not the only "*émigrés*" who contributed to the *éclat* of the Park Theatre. A French nobleman, the Baron de Rostaing, and a barrister, M. Savarin, were enabled to turn to account, both on the stage and in the orchestra, talents which in early life they had cultivated only as sources of individual gratification and social amusement.

An anecdote is related of the young architect during his connection with the theatre, illustrative not only of his ingenuity, but of his love of a joke. At a grand public masquerade given to inaugurate the opening, an elegantly constructed locomotive windmill made its appearance on the stage, the only apparent opening to which was a window near the top. The singularity of the construction excited, naturally, a surprise, which was increased to astonishment when a voice was heard to issue from the machine, uttering a variety of

D

political, as well as personal satires ; and exhibiting an intimate acquaintance with the social condition of New York. This could not be long endured. A call was made for the *Thersites* of the mill to shew himself, under a loud threat of summary chastisement by the demolition of the machine and the exposure of the *frondeur*.

When the excitement was at its height, and the destruction of the windmill seemed inevitable, the machine was gradually brought over one of the trap doors on the stage. Brunel, and the companion whose wit had led to the anticipated catastrophe, allowed themselves to drop gently through, and thus to effect their escape from the theatre undiscovered. The disappointment of those who had already breathed a vow of vengeance may be well conceived when the machine was found to be untenanted ; and as Brunel and his friend left New York that night for Philadelphia, the mystery remained unexplained.

However well calculated were the designs for this theatre to exhibit an unusual amount of talent and resource, and however the execution of them may have served as an introduction to more general architectural practice, the work failed to procure Brunel any direct pecuniary benefit. Unfortunately this building was burnt down in 1821, and there remain no authenticated drawings to shew the architectural novelties of its construction. The cupola by which it was surmounted is said to have resembled that in Paris over the Corn Market ; while in the boldness of its projection and the lightness of its construction it was far superior.

So high had Brunel's talents raised him in the estimation of the citizens of New York, that they resolved to appoint him their Chief Engineer. In that capacity he was soon called upon to prepare designs for a cannon foundry. Up to Brunel's advent no establishment of that kind existed in the State; nor does it appear that Brunel had ever directed his attention to that branch of engineering. At Douai, Ruelle, and Strasburg, the old method of *loam-moulds*, and partially hollow castings, with the subsequent application of the cutters, or *alésoirs*, for boring, still prevailed; but of this method Brunel had no practical knowledge, no more than he had of the improvements introduced into England, where, at that time, about 27,000 tons* of iron were being annually converted into cannons, mortars, carronades, shot and shells.

If, however, the want of precedent brought a greater demand on his invention, it also relieved him from the paralysing influence of authority. Left free to solve the problems presented to him, he very soon organised an establishment for casting and boring ordnance, which, from its novelty, practicability and beauty, was considered, at that time, unrivalled; and which in itself was sufficient to place its originator in the foremost rank of mechanical engineers.†

Shortly after the completion of the theatre at New York, Brunel was called upon to mourn the loss of his enlightened patron and liberal friend, M. Pharoux, *amicus usque ad aras.* He had returned to his hydraulic avocations on the Black River, one of the

---

* Board of Ordnance 11,000 tons, East India Company 6000 tons, trading and other armed vessels 10,000 tons.

† " L'ingénieux mécanisme qu'il imagina pour exécuter l'opération du forage des canons, ses nouveaux alésoirs, l'adaptation des mouvemens par le moyen desquels il remuait, il faisait tourner facilement des masses si lourdes, une foule d'inventions et d'idées fécondes qu'il mit au jour, suffiraient pour établir sa célébrité."—*Notice historique par Frère.*

most turbulent of the northern streams. This river takes its rise on the western declivity of the Essex Mountains, pursues a course of about 120 miles, sometimes interrupted by cataracts, and sometimes hurried onward by rapids, until it ultimately discharges its waters into Lake Ontario, at Sacket's Harbour. In his attempt to cross the great falls of this river, M. Pharoux, and seven of his companions, perished; a fate to which, we almost shudder to think, Brunel might have been also exposed, had not his genius and a protecting Providence opened to him another and a safer path.

New York seems to have been, at this time, considerably indebted to French genius for many of its most important works.

The defence of the entrance to its land-locked bay had long been in contemplation. Between Staten Island and Long Island the bay contracts to the width of a mile, and receives the name of " Narrows." To a French officer of talent and experience, Major L'Enfant, the projection of the defence of this channel had been confided ; but the evidences which were so fast accumulating of Brunel's engineering qualifications, scarcely justified the citizens in neglecting to secure his opinion and assistance. Accordingly, plans were obtained from him which seem to have been those ultimately adopted.

In 1796, we find Brunel admitted to the privileges of a citizen of New York. (See Appendix A.)

Of the amount and variety of his labours, and the difficulties against which he had to struggle during his residence in that city, there remains, unfortunately, no record. We have, however, incidental testimony that his genius received but inadequate reward in America; notwithstanding which, he resolutely declined to entertain urgent and repeated invitations to return to his own country. France had now entered upon a new phase of her political existence; she had shaken off the yoke of the sanguinary monsters of the Revolution, and had established an Executive Directory which afforded some guarantee for good order and wise polity; her arms were everywhere triumphant. Holland, under the title of the Batavian Republic, had become her ally; Russia had deserted her coalition with Austria; and Austria herself, by the Treaty of Campo Formio (October 17, 1797) had been compelled to acknowledge the power of France. Notwithstanding all this, Brunel felt that his country offered no real security, either for personal or political freedom. He was still apprehensive that the leaders of the Revolution were scarcely prepared to understand the true principles for which they were contending, and were, therefore, little likely to use with discretion the power with which they might become invested. He had learnt to think that freedom was of progressive growth, and that France, which had been so long deprived of the first elements of liberty, could not suddenly be brought to walk in the steps of America, when there was no sagacious and disinterested Washington to guide her councils.

Respect for constitutional authority ever formed a leading characteristic of Brunel's mind; no man more strongly condemned that disregard of moral feeling which so generally obtained in relation to the crimes of great military or political usurpers. The formal censure which our Christianity passes upon usurpation and tyranny, had for him a reality—not to say a solemnity—of conviction. It was, therefore, no wonder that he should continue to resist every temptation, even in after life, and when the imperial power had been firmly established, to take up his abode in France.   At the period to which we now refer, Brunel one day received an invitation from Major-General Hamilton — the distinguished aide-de-camp and secretary to Washington—to meet at dinner a M. Delabigarre, recently arrived from England.   The absorbing subject of conversation in all society was the triumphs achieved by the British navy at Cape St. Vincent and at Camperdown; at General Hamilton's table the naval prowess of England formed naturally an interesting matter of discussion, leading to a consideration of the principles of naval architecture and the supply of the materials of ships of war.

M. Delabigarre seemed to have directed his special attention to these subjects, enlarging more particularly on the manufacture of *ship's blocks.*

He described with accuracy the nature of the machinery in use at Southampton by the Messrs. Taylor, and spoke of the large and increasing cost of those articles.

Brunel listened with attention and with interest, pointing out what occurred to him as defects in the manipulation, and suggesting that the mortises in the shells of the blocks might be readily cut with chisels, two and three at a time.

There are no records to show how the suggestion here thrown out took root in his mind, developed into form, and ultimately expanded to proportions so great as to embrace the whole requirements of the British navy.   A memorandum in one of his subsequent journals simply states that " the shaping machine I conceived while I was roaming on the esplanade of Fort Montgomery; then not a house was in sight, except at

the landing below and at Verplante Point." If, however, Brunel laboured under the disadvantages of want
of experience and example in the manner of accomplishing his work, he was the more strongly impressed with
the necessity of giving his whole mind to the questions
presented to him. To investigate a problem upon its
own merits is not always easy. To conceive the end
to be accomplished requires a mind more comprehensive in its grasp, and is therefore more rare in its development, than that which exercises itself in the means
to be employed ; but when both the imaginative and
constructive faculties are happily united, we have the
real inventor, him to whom antiquity accorded the
highest honours.

" Founders and senators of states and cities, lawgivers,
extirpers of tyrants, fathers of the people, and other
eminent persons in civil government," says Lord
Bacon, " were honoured but with titles of *worthies* or
demi-gods ; whereas, such as were inventors and au-
thors of new arts, endowments and commodities
towards man's life, were ever consecrated amongst the
gods themselves : and justly, for the merit of the former
is confined within a circle of an age or a nation, and
is like fruitful showers, which, though they be profitable and good, yet serve but for that season, and for a
latitude of ground where they fall ; but the other is,
indeed, like the benefits of Heaven, which are permanent and universal, coming ' in aurâ leni,' without
noise or agitation." *

The influence of authority, which education is too
well calculated to produce, while it tends to remove
the asperities and smooth the irregularities which in-

* Advancement of Learning.

terrupt and sometimes endanger the interests of society, has the effect, also, of repressing originality of thought and of weakening the faculty of invention. "Men there have been," says Macaulay, "ignorant of letters; without wit, without eloquence; who yet had the wisdom to devise and the courage to perform that which they lacked language to explain. Such men have worked the deliverance of nations and their own greatness. Their hearts are their books; events are their tutors; great actions are their eloquence."

"Les grands services font les grands hommes, car la vraie gloire n'appartient qu'aux idées fécondes."

CHAPTER IV.

1799–1801.

QUITS AMERICA — DUKE OF ORLEANS — LANDS AT FALMOUTH — PREJUDICE AGAINST FOREIGNERS — JOHN FELTHAM — MACHINE FOR TWISTING COTTON AND FORMING IT INTO BALLS — MACHINE FOR HEMMING AND STITCHING — MACHINE FOR CARD SHUFFLING — DESIGNS FOR A BLOCK MACHINERY — EDWARD, LORD DUDLEY — SLIDE REST — MAUDSLAY — DESIGNS OFFERED TO MR. TAYLOR REJECTED — INCREASE OF THE NAVY — SIR SAMUEL BENTHAM.

ON the 20th January, 1799, Brunel bade adieu to America, grateful for the freedom which her institutions had permitted him to enjoy; for the encouragement which her citizens had afforded to his expanding genius; and for the opportunity which the requirements of its rudimentary condition offered for testing the practical value of his projections.

The hope of his boyhood, once to visit England, and which found expression on the quay of Rouen, when he exclaimed, " Ah! quand je serai grand, j'irai voir ce pays-là," was at length to be fulfilled; and the unwavering constancy of a long-cherished attachment was now to meet with its reward.

The reputation which he had acquired in America enabled him to procure valuable letters of introduction to individuals in this country, from men of learning and eminence in the United States. With H. R. H. the Duke of Kent he had already formed a personal acquaintance in New York, where he also became

known to the Duke of Orleans, afterwards King Louis
Philippe; and who, at a subsequent period, when en-
tertaining Brunel at the Palais Royal, seemed pleased
to recall the circumstances connected with their first
acquaintance. He would cheerfully remind Brunel
how he and his brothers, the Duke de Montpensier and
the Count Beaujolais, fared; sometimes compelled to
quit the only inn in a wild district in consequence of
some unintentional offence offered to the landlord, and
sometimes compelled to perform long journeys on foot,
each with his luggage on his back. " Ah ! " said his
Majesty, " c'était vous Brunel qui y voyageait en grand
Seigneur; but I with my friends went through that
country very differently; obliged to support ourselves
principally by our rifles,— the clouds of heaven our
canopy, the trees of the forest our bed-curtains."

In March 1799, Brunel landed at Falmouth, and
shortly after was united to Miss Sophia Kingdom.

It was now no longer the loved image only which,
in the anxious yearning days of youthful exile, his
faithful pencil had so often embodied; but the living,
confiding woman that he now pressed to his heart, in
all the flower of life;— she who had, for his sake,
steadily rejected many an eligible suitor which her
fascination and beauty had attracted. We may well
believe that her confidence and affection had nothing
to regret when, with the dew of youth upon his heart,
and the smile of truth upon his lips, he could write in his
seventy-sixth year, and after forty-six years of wedded
life, this touching acknowledgment, that " To you,
my *dearest* Sophia, I am indebted for all my success."
Truly says Jeremy Taylor, " She that is loved is
safe, and he that loves is joyful."

Necessity no more than inclination would permit

Brunel to remain long unemployed, although a serious doubt might be well entertained as to the reception which would be accorded to him by the country.

The hereditary feeling of repugnance to everything French, and indeed the suspicion and jealousy with which everything foreign was regarded, attained in England an intensity, about that period, greater perhaps than at any other on record. The length to which these feelings were carried is curiously illustrated in a letter placed in my hands by Sir Benjamin Hawes, K. C. B., and addressed by Mr. John Feltham, a gentleman well known and respected, to Sir Benjamin's father.

" Bath, No. 12, Kingsmead Square,
" 20th July, 1798.

"Dear Sir,

" You need not be alarmed when I tell you I am in custody, and my papers, writings, &c. &c., seized by order, I apprehend, of His Grace the Duke of Portland, in consequence of having a few weeks since given a poor Turk, or Persian, 2s. 6d. and a breakfast. I could not understand a word he said ; but he was poor, and on foot. He brought me a letter of recommendation from Mr. Hoskins, who had, I suppose, relieved him ; and he desired me to copy a letter to give him to take on to London to Mr. Wilmot to give him 10s. 6d. in the Boro', — which letter has been seized, and the poor man taken up for a spy.

" This letter of Mr. Hoskins, which I copied, mentioned the word ' citizen,' which it seems has caused the alarm. I have no doubt of the business proving highly honourable to Mr. H. and myself : though I may possibly be brought to town. I underwent a long examination yesterday, and the Bench and Mayor behaved very much to their credit, and paid me some compliments for my conduct. I sent for Mr. Cruttwell, on whose word I am on the parole of honour.

* * * * * * * * *

"Your affectionate friend,
"JOHN FELTHAM.

"P.S.—Probably this little act of benevolence has stopt all my letters. Adieu. God bless you.

"To Mr. Benjn. Hawes."

Brunel had indeed introduced his inventions from America, and it was at first supposed that they were of American origin, which may possibly have had the effect of modifying feeling and opinion. Still he was unable to enter the gates of Portsmouth Dockyard without an official order or permit, even when engaged in superintending his own works.

In May 1799, Brunel took out his first patent. This was for a duplicate writing and drawing machine. In principle it resembled the Pantograph, as described in the "Mémoires de l'Académie des Sciences," 1743, though differing widely in the details. A machine for twisting cotton-thread and forming it into balls was also amongst the earliest of Brunel's inventions in this country. The impulse given by this machine to the employment of cotton can now scarcely be credited. The little balls were very elegant in form; and from the manner in which the thread was wound, they presented the appearance of net-work, or ribbons of lace. The machine measured the length of the thread which it wound, and proportioned the size of the ball to its weight and fineness. Unfortunately Brunel neglected to secure the benefit of his invention by patent, and it was therefore rapidly and generally adopted; and while thousands of pounds were realised through its means, Brunel himself remained without remuneration. In his Journal of 1806, he notices a visit which he paid to the establishment of the Messrs. Strutt, at Belper (Derby), where, after remarking that there were 640 persons employed, he says, " I observed they had adopted my contrivance for winding cotton into balls. There were about twenty spindles on one swing." A lady, a friend of Brunel, having experienced the advantage of the little cotton balls, while expressing her admiration to him, jokingly suggested that he ought to invent a means of relieving ladies from the wearisome employment of hemming and stitching. To any other, the observation would have passed as it was intended. It was certainly forgotten by the lady herself ; when, to her surprise, his patent for "trimmings and borders for muslins, lawns, and cambric " was shown to her, and in which she found her wishes more than fulfilled. The advantages of this invention are stated to be " that the operations of hemming, whipping, or otherwise securing from ravelling the edges of trimmings cut in narrow slips out of border webs, as they have unavoidably been hitherto, are by this invention altogether saved." To this machine may perhaps be referred the origin of that recently introduced from America, and so largely employed in Belfast and the north of Ireland in hemming cambric handkerchiefs, stitching linen drawers and jackets, and in making shirts.

A very essential difference will be observed in the fate of the two machines. While the one remained neglected and unproductive, the other is a marked success, and the object of an important and remunerative trade.

Brunel also invented, about this period, for the benefit of some feeble-handed card-player, a little machine for shuffling cards; but what the exact nature of its construction was, I have been unable to learn. The cards were placed in a box, a handle was turned, and in a few seconds the sides of the box opened, presenting the pack divided into four parts, and the cards most effectually mixed. This machine he presented to Lady Spencer.

In a note addressed to Lady Hawes by the Dowager Lady Littleton, she says that "she well recollects Sir Isambard bringing to her mother the little instrument for shuffling cards,—and also the deep interest and admiration with which her parents always thought and spoke of him."

Although these machines afforded no direct profit, they served as an introduction, and as offering valuable testimony of Brunel's mechanical genius and skill. It was, however, to a *system* of machines which should supply the whole British navy with blocks, that we must look for the establishment of his claim to occupy the first rank amongst inventors and mechanists. Owing to the fortuitous circumstance of Mrs. Brunel's brother being Under-Secretary to the Navy Board, Brunel was enabled, through him, to enter into negotiations with Messrs. Fox and Taylor, who had for many years enjoyed a monopoly for the supply of blocks to the British navy, and to whom Brunel first made offer of his ingenious inventions,—with what result we shall presently see.

So long ago as 1775, Mr. Taylor took out a patent " for the improvement in coghing or bushing of cast iron or metal shivers for ships' blocks;" and in 1781, another patent for " planking shivers with lignum vitæ, or other hard wood, so that the pieces of plank let in on each side of the shivers to cross each other shall wear on the pin head or endway of the grain with little wear, and less noise or friction than heretofore." This patent included the bushing, boxing, coghing, or plating the shivers with hard wood for forming the rim or groove of shivers in cast metal ; the shivers to have spokes of lignum vitæ, or other hard wood, and to be secured to the rim by screws or rivets ; and for boiling English wood shivers in oil or salt water " to render them more serviceable."

Still the most essential operations connected with block-making were performed by manual labour ; and upon the accuracy of the eye and hand of the workman depended the execution of the work. This was, however, not only highly costly, but, from want of uniformity in the execution, disappointing and unsatisfactory. The difficulties which stood in the way of Watt might have postponed also the accomplishment of Brunel's project, had not the kindred inventive genius of Henry Maudslay supplied a highly important element of success ; and thus has it ever been that ideas are found to precede, sometimes for years, their practical fulfilment.

"The principle of the press," for example, " which bears the name of Bramah, was known," says Mr. Babbage, " about a century and a half before the machine to which it gave rise existed ; but the imperfect state of mechanical art in the time of the discoverer would have effectually deterred him, if the application of it had occurred to his mind, from attempting to employ it in practice, as an instrument for exerting force."*   In another department we have

---

* History of Machinery and Manufactures, by Charles Babbage, F.R.S.

an example in the ruin to which Edward, Lord Dudley, was exposed, when, in 1619, he sought to realise his idea of applying pit coal in place of wood fuel to the smelting of iron ore, a process which, beyond all other, has gained for England her superiority in the mechanical arts, but which the ignorance and the prejudice of the people rejected for nearly one hundred years.

In the case of Brunel, his mechanical conceptions could scarcely have been developed without the aid of the *slide rest.*

Mr. Nasmyth has shewn*, that by the application of this instrument to the turning lathe, the whole condition of practical mechanism was .changed. " A mechanical contrivance was made to take the place of the human hand for holding, applying, and directing the motions of a cutting tool to the surface of the work to be cut, by which the tool is constrained to move along or across the surface of the object with such *absolute precision,* that with scarce any expenditure of force on the part of the workman, any figure, bounded by a line, a plane, a circle, a cylinder, a cone or a sphere, may be executed with a degree of accuracy, ease and rapidity, which, as compared with the old imperfect *hand system,* may well be considered a mighty triumph over matter." " It is to this instrument," says Mr. Nasmyth, " we owe the power of operating alike on the most ponderous or the most delicate pieces of machinery, with a degree of minute precision of which language cannot convey an idea."

In estimating then the value of the vast change

---

* Essay on Tools and Machines, appended to Buchanan's Mill Work, revised by George Rennie.

created in our dockyards by the genius of Brunel, we must not forget how much is due to the able co-operation of Mr. Henry Maudslay, who, by a happy combination of industry and genius, was enabled, from the humblest beginning, to build up an establishment which gave to the world such men as Field, Nasmyth, and Whitworth to perpetuate its character, and to confer upon their country the highest mechanical benefit yet obtained. Perhaps no nobler monument has been raised to the invention, the skill, and the perseverance of an individual mind than that now exhibited in Cheltenham Place, Lambeth.

When, in 1800, Maudslay was engaged upon the working model of the block machinery in Wells Street, one assistant was found all-sufficient for his wants. In those workshops, which he, in conjunction with his respected, now venerable, partner, Mr. Field, established in Lambeth, there may be now seen upwards of 1200 mechanics, many of superior attainments and skill, carrying out some of those vast engineering appliances which the requirements of the country demand, and in the preparation of which some of the original constructions, if not inventions, of Maudslay, may be still seen to take an important part.

The skill of Mr. Maudslay became first known to Brunel through a M. de Bacquancourt, a French emigrant of considerable mechanical dexterity, and who, by some happy accident, had made the acquaintance of Maudslay. Between M. de Bacquancourt and Brunel there was a natural mechanical sympathy; but the disposition and the politics of M. de Bacquancourt, which were of the ultra Royalist stamp, prevented any very intimate connexion.

By the early part of the year 1800, Brunel had not

E

only completed his drawings of the principal parts of the block machinery, but had made a working model of the mortising and boring engines, it is believed principally with his own hands, which left no doubt as to the practical value of his projection, and for which in 1801 he took out a patent.

Under the impression that the contract still subsisting with Messrs. Fox and Taylor for supplying blocks to the navy would present a serious obstacle to the introduction of his machinery, he naturally made, through Mr. Kingdom, an offer to those gentlemen of the results of his labours.

The reply of Mr. Samuel Taylor is as follows :—

" Southampton, March 5th, 1801.
" Dear Kingdom,

" I am favoured with your letter of the 2nd inst., and I should have replied yesterday but I had not time.

" Your brother has certainly given proofs of great ingenuity, but he certainly is not acquainted with our mode of work.   What he saw at Deptford is not as we work here.   I will just describe in a few words how we have made our blocks for upwards of twenty-five years — twenty years to my own knowledge.   The tree of timber, from two to five loads' measurement, is drawn by the machine under the saw, where it is cut to its proper length.   It is then removed to a round saw where the piece cut off is completely shaped, and only requiring to be turned under the saw.   The one, two, or three, or four mortises are cut in by hand, which wholly completes the block, except with a broad chisel cutting out the roughness of the teeth of the saw, and the scores for the strapping of the rope.   Every block we make (except more than four machines can make) is done in this way, and with great truth and exactness.

The shivers are wholly done by the engines, very little labour is employed about our works, except the removing the things from one place to another.

" My father has spent many hundreds a year to get the best mode, and most accurate, of making the blocks, and he certainly succeeded; and so much so, that *I have no hope of anything ever better being discovered, and I am convinced there cannot.* At the present time, were we ever so inclined, we could not attempt any alteration. We are, as you know, so much pressed, and especially as the machine your brother-in-law has invented is wholly yet untried. Inventions of this kind are always so different in a model and in actual work.

" Believe me, dear Kingdom,
" Yours in great truth,
" SAMUEL TAYLOR."

I may here mention that the average supply of blocks during 1797-98-99, 1800 and 1801 was 100,000, the value of which, with the other articles of the block-maker's contract, amounted to about 34,000*l.*

In 1793 there were 153 ships of the line, and 411 below that rank. In 1803 there were 189 ships of the line, and 781 below that rank. The tonnage had increased from 402,555 to 650,976, or 61 per cent.* It was no wonder then that the contractors for blocks were overpowered with work, and that to meet the increasing demand some change was required to increase the supply, and check the cost.

Brunel had now no hope of inducing the con-

* According to the Report of the Surveyor-General of Land Revenue and Roads and Forests, the Navy had increased in 1806 to 776,087 tons.

tractors to adopt his inventions ; but having discovered that the valuable monopoly which they had so
long enjoyed was about to expire, it became of the
utmost importance that he should obtain an opportunity
of laying his invention before the Government authorities.

Fortunately for Brunel, and for the country, Lord
Spencer had not yet left the Admiralty ; to him Brunel
had brought an introduction from America, and now he
was by Lady Spencer made known to Sir Samuel
Bentham, K. S. G., who filled the important office of
Inspector-General of Naval Works. This office had
been specially created for Sir Samuel, that he might be
freed from the control of the Navy Board, a govern-
mental department to which the execution of works
determined upon by the Admiralty was confided, but
which seems to have been only calculated to enlarge
patronage, decrease responsibility, and multiply the
links in the official drag-chain of the naval service.

It must be remembered that it was only under the
administrations of Lord Spencer and Lord St. Vincent
that for many years any improvement had been attempted in the naval department. The difficulties with
which those men had to contend in overcoming the
force of inertia and the spirit of routine, which from
the end of the seventeenth to the commencement of the
nineteenth century pervaded our naval administration,
must have been enormous.

The steam-engine had from the commencement of
the eighteenth century been applied to our large
mining operations with increasing advantage. It had
also, from 1780, rendered immense service to the
manufacturing interests of the country ; yet, until 1798,
it had been excluded from our dockyards.

It was the same with regard to the application of

machinery to the manufacture of cordage, anchors and blocks. Of this neglect no one was so conscious as Sir Samuel Bentham, and no one laboured more diligently than he to bring about the necessary reforms. With regard to the special improvement now proposed by Brunel, the very position occupied by Bentham might have proved the greatest impediment to its success. Bentham was himself an inventor and mechanist of the highest distinction. He had already conceived a system of machinery for making blocks. His name was known and his influence had been felt throughout Russia. A personal friend of the Empress Catherine, he had been employed by her in a variety of important works, the Fontanka canal, the manufactories at Kritschev, the arsenal at Cherson, &c. He had been appointed *Conseiller de la cour;* he had received military rank, a gold-hilted sword, and above all, the cross of the order of St. George * ; still these honours induced no relaxation in his intellectual labours. The continuous efforts during fifty-seven years to realise his enlarged mechanical conceptions, show how deeply his mind was impressed with the importance of substituting machinery for the " uncertain dexterity of more expensive manual labour." It appears that, from 1773 to 1830, Sir Samuel's mind was constantly engaged in projections of mechanical utility, many of them anticipating the requirements of the age ; some only now adopted without acknowledgment, and some remaining still to be applied.

It was but natural to presume, that a mind so well impressed with its own superiority, as Bentham's might have been, would hesitate to admit the claims of a rival : but Sir Samuel's mind was not cast in a common

---

* Memoirs of Sir Samuel Bentham by his widow.

mould. Rising far above professional vanity and official jealousy, and consulting only his country's benefit, he no sooner became satisfied of the superiority of Brunel's inventions than he at once abandoned his own less perfect conceptions, and with a candour worthy of all praise, he did not delay an hour to forward Brunel's application to the Admiralty ; thus seeking in a noble and generous spirit to reflect upon French genius some of that honour and protection which he had himself experienced when a sojourner in a foreign land.

It is much to be regretted that no detailed memoir exists of the life of this remarkable man ; that published by his widow only whets the edge of our desire to know more of his inner self : of his trials as well as of his triumphs. From what we do know, a curious and interesting parallel is suggested between him and his protégé ;—both so largely endowed with mechanical aptitudes ; both commencing their public career far from their native land, and unsupported by the patriot's ardour ; both putting forth their best energies under the influence of their first attachments ; and both closing their missions as the active and distinguished supporters of that mechanical progress, which is ever found to be so intimately connected with national superiority.

CHAPTER V.

BRUNEL'S CLAIMS TO BE THE AUTHOR OF THE BLOCK MACHINERY
VINDICATED.

BEFORE we continue the thread of our narrative, we are called upon to encounter claims of a most uncompromising character, which have been, I believe, for the first time set up on behalf of Sir Samuel Bentham to the authorship of the block machinery. In the "Mechanic's Magazine" (April 3, 1852) assertions have been made which would have the effect of wresting from Brunel the honour which is his due, and of bestowing it upon one whose nobleness of mind and disinterestedness of character never permitted him, while he lived, for one moment to appropriate to himself the inventions of others.

The effort to render justice to the emanations of original minds engaged in similar contemplations, and to assign to each the exact amount of merit due, is not always easy. Conflicting opinions are sometimes so nicely balanced, as almost to defy our industry and our penetration in arriving at a right conclusion ; and we are, therefore, bound, from our own liability to err, to deal tenderly with any honest expression of opinion which may not entirely coincide with our own. Should it, however, be found, in the progress of investigation, that personal or partisan feeling had been permitted to usurp the place of patient and discriminating inquiry, leading to conclusions not consistent with the evi-

dence, then we are equally bound to exhibit the
fallacy, and to dissipate the illusion to which such
feelings inevitably give rise.

The claim is comprehensive and unqualified ; and is
repeated in Adcock's " Engineer's Pocket Book " for
1856. It includes " all the operations preparatory to
the shaping of blocks, with the pleasing and convenient arrangement of the block machinery, whereby a
regular sequence of operations is obtained ;" while it
allows to Brunel " some few only of the operations
requisite for the shaping and finishing the blocks ;"
although, " in these instances, the means of performing
the requisite operations were rarely other than those
specified in Bentham's patents."

According to this view, Brunel can be regarded
only as draftsman or clerk of works to Bentham.
This opinion, however, is in a subsequent paragraph so
far modified as to allow " Brunel some share in the
arrangement."—" It cannot be supposed that Bentham
contrived every detail ; that was Goodrich's particular
duty, and Brunel had his share in the arrangements ;
sometimes advantageously, at others introducing wheels
that would not work, as appears from a pencil sketch
now lying on the table."

Without desiring to analyse too critically the fore-
going paragraph, we shall only observe that *detail* is of
the utmost importance to the success of constructions
altogether new ; and so strongly was Brunel impressed
with the necessity for this, that as we have seen he
depended neither upon a *pencil sketch*, nor a finished
drawing ; but was willing to incur the expense of
working models where any difficulty was likely to arise.
Without knowledge it is easy to condemn.

When in 1807, and during Bentham's absence from

the country, a new engine was introduced at Portsmouth for cutting copper bolts, the master shipwright informed Brunel that it "was found incapable of performing its operation." Brunel at once proceeded to inspect it. " I observed," says he, in his journal, " that the saw, or circular cutter, had no set; that the moving frame was unmanageable; and that the manner of laying the bolts was imperfect.

" I gave directions to Barlow (Maudslay's mechanist) to put a counterpoise to support the weight of the swinging or moving frame, and to apply to the saw a proper set.

" At half-past four o'clock in the afternoon I had it tried before the officers; *it produced its effect with the greatest celerity.*"

Note.—" Such a machine ought to have been accompanied by a person acquainted with the use of it."

But this wholesale advocacy of the claims of Bentham, at the expense of Brunel, can scarcely be considered as either becoming, or wise, where evidence was at hand, sufficient, at all events, to induce a candid mind to hesitate. Without justice there can be no honesty. *Nihil honestum esse potest quod justitiâ vacat.*

It is asserted, that " Brunel's *drawing* was at that time (1802) confined to the shaping of a block shell; while Bentham's *machines were already in the dockyard in a working state.*" The italics are the writer's. They intimate very distinctly that Brunel had nothing but a *drawing* to offer in illustration of his project, and that *that drawing* was limited to a description of his *intended* mode of shaping a block shell. They also intimate that Bentham's machines were already in operation. But we shall find that neither of these assertions has any foundation in truth.

We have already seen that before Brunel had obtained an introduction to Bentham—before, in fact, he had negotiated with the contractors, who were at the time "so much pressed that they could undertake no alteration in their system "— he had already prepared *working models* of two of the most important engines : viz. those for *mortising* and *boring*. But the simple statement now before me in Brunel's own handwriting —obviously never intended for publication—shows what his real position was, with regard to Bentham.

Being so far disappointed [in his application to the contractors] " I intimated my intention of exhibiting the plans and models to General Bentham, who had in contemplation at that time the formation of a block-making establishment from machinery of his own. The steam engine was already up in Portsmouth dock-yard, and the building very far advanced. The steam engine was coupled with the pumps that were destined for the occasional service of draining the docks ; ample share of its power could therefore be employed in making the blocks for the navy, for which General Bentham had already made his arrangements and some of its parts. At the change of the administration to that of Lord St. Vincent, I had an opportunity of submitting my plan to General Bentham, who came to see the chief parts.

" On the production of the results, he admitted *at once* that I had left no room for any part of his plan, and promised to make the admission to the Lords of the Admiralty." In consequence of this candid expression of opinion, Brunel addressed a letter, February, 1802, to Mr. Evan Nepean, Secretary to the Admiralty, of which the following is a copy :—

Sir,

" Having been informed that it is the intention of Government to have the blocks for the use of the Royal Navy manufactured in H.M. dockyard at Portsmouth by means of steam engines already erected there, and that workhouses for the reception and accommodation of machinery for that purpose are now erecting, it occurs to me as not an improper opportunity for requesting permission to submit, through you, for their Lordships' consideration, the following tenders :

" I have invented and executed new engines, by the operation of which blocks may be manufactured with infinitely more celerity and exactness than they can be done by the machines at present in use.

" I beg leave to represent that these engines cut the mortises, and shape the outside of the shells in such manner, that without requiring dexterity on the part of the workman, the shells of a determined fixed size cannot differ one from another, either in the proportion of the mortises, or in the shape and dimensions of the outside.

" The inconveniences to which blocks are constantly liable by the friction of the cords against one or alternately both sides of the mortises, are remedied by introducing a sheet of metal, bent to the shape of the upper part of the mortise.   This operation is also performed by a particular-engine.

" The shivers, with metal coaks made by these engines, are executed with precision and celerity, and any number of a determined size being gauged most minutely, it will be found that one does not differ from another, either in diameter or thickness, so that any one

of these shivers will suit equally well any shell of the
size for which it was intended.

" The engines, which I am the inventor of, extend
no farther than for the making of the shells and
shivers.   The pins, either wood or iron, I propose
should be made in the manner already in use.

" The advantages which result from the use of
these engines, consist in obtaining *uniformity*, *exactness*, and *celerity*, without relying on the dexterity
of the workmen; and owing to the peculiar principle on which the blocks are shaped, they cannot
be counterfeited — a circumstance to prevent embez-
zlement.

" *I have executed a working model* which I should
be happy to have the honour of submitting to your
inspection, and will send it to the Admiralty any day
you will please to appoint.

" I have the honour to be,

" Your obedient servant,

" I. BRUNEL."

The Italics are ours.

Brunel goes on to state in his journal;—" A few
days after, I received an order to be at the Admiralty
with my small models, which gave such satisfaction,
that my proposition of making a block mill was
adopted.   Accordingly General B. took me to Portsmouth.   Having had occasion then of seeing what
had already been done of the steam engine and buildings, I made my disposition accordingly.   But a
most difficult task was to find some person fit for the
execution of so extensive and so complicated an apparatus.

" So backward was the mechanical industry at the
time that even *English wrought iron* was prohibited

from all Government supplies, and the cast-iron was of too brittle a nature for general use."

So far then from Bentham's machines being in a "working state," nothing was found at Portsmouth but the steam engine and some buildings.

On the 2nd April, 1802, Brunel addressed the following letter to Sir Evan Nepean, Bt. :—

" Sir,

" Their Lordships having appeared satisfied with the model for making blocks which I had the honour of showing to them, I feel anxious to know whether it is their Lordships' intention to have them established for the use of H. Majesty's Navy.

" The advantages which would be derived from an establishment of that kind, could not be pointed out or ascertained from a small model, but the machines *I have* executed on larger proportions, have, by their produce, enabled me to make the following estimates of the prices at which the various sizes of blocks could be manufactured.

" The blocks made with the assistance of my machines are executed with exactness and expedition, which will afford a considerable saving on the present cost, exclusive of the advantage of employing a great quantity of wood which is wasted in the yards.

" The following estimate made on four sizes only, namely, eight inches, twelve, sixteen, and twenty-one, will evince the proof of what I asserted :—

	8 In. s. d.	12 In. s. d.	16 In. s. d.	21 In. s. d.
Prices at which blocks may be made	1 8¾	4 5	8 11½	18 1¾
Prices allowed by Government .	2 3½	6 0½	13 6	27 0½
Saving	0 6¾	1 7½	4 6½	8 10¾

" I have the honour to be, &c."

" The merit of the few operations requisite for the shaping and finishing of blocks," and the questionable share in " the arrangements " which had been, at first, conceded to Brunel by the writer in the " Mechanic's Magazine," is subsequently sought to be effaced, by the institution of an invidious comparison between the limited purposes to which those operations could be applied, and the *general* applicability of Bentham's machines; and a report, dated November 1804, is quoted, addressed by Mr. Samuel Goodrich, described as Bentham's mechanist, to his principal, from which a strange inference is obtained.

The report is as follows :—

" None of the existing machinery, more immediately belonging to the mortising, shaping, and boring of the shells of blocks, can be well applied to any other purpose as far as appears at present.

"The circular saws, and up-and-down saws, *can* be applied to general purposes, and others may be introduced for cutting."

The arbitrary and anomalous conclusion drawn is, " That the above communication seems in itself sufficient proof, that great part of the machinery comprised in that for block-making was, from the first, and still continues to be, of Bentham's invention—not of Brunel's." Not only, then, is Brunel denied the honour of being an inventor, but also the credit of being a manipulator, and is at once degraded to the position of a plagiarist —an adapter of other men's designs, " having Bentham's patents before him," and every " opportunity of seeing the Bentham machinery in Queen's Square Place, and having farther secured the mechanical skill of Henry Maudslay, *who made* the machine after frequent consultation with Bentham, and examination of them often whilst in progress of manufacture. Bentham, Goodrich, Burr (superintendent and draughtsman)—all of them discussed the suitableness for its destined work of every particular engine, each of them indicating means by which it might be more or less improved."

It must be obvious that this statement proves too much, and implies that the block machinery, in place of being a beautiful and symmetrical system, emanating from one mind, being directed to one end, and consistent in all its parts, was but the result of a heterogeneous concatenation of masters, mechanists, and inventors. The only machines of Bentham's construction really applicable to block-making were saws ; but they were soon found quite inadequate to fulfil the duty which the block-machinery of Brunel demanded.

" The imperfection of the various mechanical contrivances that had been invented for the purpose of sawing, led me," says Brunel in one of his communications to the Admiralty, "to direct my views to the invention of such machinery as should be the means of obviating these difficulties, and I foresaw that a field would be opened to me of rendering service to the naval establishments of the kingdom, of a magnitude much exceeding those which had been derived from my improved system of making blocks."

It appears that Mr. Burr, Bentham's draughtsman, had been appointed superintendent of the wood mill at Brunel's request.   In October 1803, Bentham having expressed a wish that some experiment should be made with one of the saws, Brunel replies :

" I am fearful that the means used for cutting lignum vitæ cannot be adapted to advantage for common wood.   I will however try it.   Mr. Burr is solicitous to bring forward any of *your* inventions; he will readily

assist me to make the experiment." The italics are
ours.

Had any of the block-machinery, properly so called,
been due to Bentham, Brunel would scarcely have ventured to draw the distinction which he did between
the value of *meum* and *tuum* in this as in other
communications.

It is further stated that " Brunel's machines were
never sanctioned till drawings of them had been well
considered and approved by Bentham." Official position
is thus confounded with mechanical originality; and
an inference is erroneously drawn, that Brunel could
have only acted in intellectual subordination to Bentham; but any person at all conversant with official
arrangements must be aware that every order, before it
can be executed, must receive the sanction and signature
of the head of the department from whence it issues.
It has, however, been already shown, that not only
were those *drawings* in existence before Brunel had
even obtained an introduction to Bentham, and therefore entirely independent of Bentham's consideration,
suggestion, or improvement; but *models and machines,
on an effectively working scale*, had been also executed,
from which results were obtained sufficient to justify
the Admiralty in adopting Brunel's inventions. In
this discussion the authority of Dr. Rees cannot be
overlooked. He was the editor of the Cyclopædia
commenced in 1800 and completed in 1819, as an
expansion of Chambers' work published in 1786. In
that justly celebrated publication there are not less
than sixteen pages devoted to the description, and seven
admirably executed plates in illustration of the block
machinery; because as the writer observes*, " they

---

* Mr. John Farey, an engineer of recognised standing and
eminence.

are the most ingenious and complete system of machinery for forming articles in wood of any this kingdom can produce ; " and " *these machines*," he distinctly records, " *are the invention of Mark Isambard Brunel.*" Now, as this work must have been in the hands of every scientific man who could borrow or purchase it, the question naturally arises how it happened that a generation (thirty-three years) should have been permitted to elapse without any doubt or discredit having been cast on that article.

I will now add the testimony of Sir Samuel Bentham himself. In his " *Statement of Services relative to Improvement of Manufactures requisite in Naval Arsenals,*" he defends himself from the accusation of having afforded encouragement to foreign talent by appealing to the result.

" Machinery has been applied, in the introduction of which I have been instrumental," he says, " that performs nearly all the operations requisite in making blocks, by which they are better made than heretofore," and " a saving to the country is effected of not less than 16,631*l.* per annum." . . . " Their lordships having determined that *Mr. Brunel's machinery in question* should be introduced in the manner I had suggested, Mr. Brunel employed himself in the perfecting of *his machinery* in the adapting it to the particular demands of the navy, and, in concert with the mechanist (Mr. Goodrich), in contriving the best mode of *putting it up* at Portsmouth." Sir Samuel, in concluding his vindication, says, " I cannot, therefore, but feel myself justified in having recommended the engaging Mr. Brunel's services." And in allusion to the remuneration to be given to Brunel, he thus writes : " Nor is it likely that any other mode of remuneration would have rendered

F

these services so beneficial to the public as the particular one I recommended, which combined his interest so intimately with that of the public."

Here I might be well content to leave this controversy; but as there remains still amongst us one who has borne no inconsiderable part in the mechanical movements of the age, I avail myself of his valuable testimony, and thus, once and for all, terminate, I trust, this painful and unseemly discussion.

Mr. Joshua Field, the venerable and respected mechanist, has assured me, that being engaged as mechanical draftsman at Portsmouth Dockyard when Mr. Brunel was introduced by Sir Samuel Bentham, he has a perfect recollection of the condition in which the works were at that time, together with what was proposed to be done by Sir Samuel; and as he was transferred to General Bentham's office at the Admiralty in 1804, and subsequently joined Mr. Maudslay in 1805, his evidence covers the whole period of the execution of the block machinery. His statement to me is as follows: "The works in progress, when Mr. Brunel arrived were a new steam engine and some buildings *intended* for the reception of machinery, which General Bentham *had proposed* to erect, but *had not erected*.

"The General had already introduced saws of various kinds, and machines for tongueing, grooving, and rabbetting timber; but there was no *machinery whatever* especially applicable to block making.

"That was altogether the invention of Mr. Brunel. The character of the drawings was different from any we ever had before — the proportion of the parts — the whole thing, in short; and I never once heard, during all the time of my connection with the dockyard, with General Bentham, with Mr. Goodrich, and with Mr. Maudslay, that any one ventured to deny Mr. Brunel's claims to be the *sole inventor of the block machinery.*"

I would gladly have been relieved from the necessity of this long, and to some, I fear, tedious vindication ; but it would not have been possible entirely to ignore the existence of claims so unguardedly put forward by those who may be supposed capable of influencing, to some extent, the opinion of the mechanical world, and which, if admitted, would tend to rob Brunel, not only of his position as one of the first mechanists of the age, but of the yet higher privilege of occupying a place amongst the real benefactors of this his adopted country.

CHAPTER VI.

1802–1803.

DESIGNS FOR BLOCK MACHINERY ADOPTED BY GOVERNMENT —
QUESTION OF REMUNERATION — REFERRED TO SIR SAMUEL BEN-
THAM — ACCEPTED BY THE ADMIRALTY.

THE Government having consented to adopt Brunel's plans for a block machinery, it became necessary to determine the nature and extent of the remuneration to which Brunel should become entitled. With this view, the following letter was addressed by him to the Lords of the Admiralty, in June, 1802 : —

" I beg to inform their Lordships that the invention and execution of models, and of machines on a large scale, have been attended with considerable expense and labour ; and that my time for the *two last years,* has been almost entirely employed in bringing them to their present state of perfection. I will trust to their Lordships' liberality to decide on the remuneration which they may deem adequate to the merit of the discovery.

" I hope their Lordships will take into consideration the time I have bestowed in making the several draw-ings which I have had the honour to lay before them ; and that the whole of my time will become indis-pensable in surveying and directing the execution of the machines till they are entirely completed."

To this no immediate reply appears to have been

made. The question was referred to Sir Samuel Bentham, who, in April 1803, sets forth at full length his views in the following communication addresssed to Sir Evan Nepean.

"Admiralty, April 30th, 1803.
"Inspector-General's Office.

" Sir,

" In answer to yours of the 7th instant, enclosing a letter from Mr. Brunel, soliciting some remuneration for the labour and expense which he has been at, in the invention and perfecting of his machinery for the manufacture of blocks for the use of His Majesty's Navy, and signifying the commands of my Lords of the Admiralty, that I should consider and report my opinion of what may be proper to be done in the subject of that application ; I would beg leave to state, for their Lordships' information, that having examined the several articles of the machinery in question, and having seen them at work, I am fully satisfied that they are adequate to the making blocks more perfect in regard to accuracy and uniformity of shape, as well as at a much cheaper rate, than they could be made by any other means hitherto in use.

" As to the particular blocks which Mr. Brunel has sent as specimens for their Lordships' inspection, they appear from their form and the proportion of their parts, to be better suited to their intended purpose, than the blocks in general use ; but although the particular form which Mr. Brunel has adopted in the first instance, should, after farther consideration or experience, be deemed anywise objectionable, the engines could, on any day, be set to any other form or proportions, which may be decided on as preferable ; and whatever that form or those proportions may be, there

will be no doubt but that the blocks manufactured by
these engines will, every one of them, be made in
future of that exact form, until there be found reason
to change it.

" In regard to the compensation to which Mr. Brunel
may seem entitled for the invention of these machines,
considering the great ingenuity displayed in this invention, the length of time which it must have required
to bring such an apparatus to its present state of perfection ; and considering that although the saving of
expense in the manufacturing of the article in question
is the principal object of the invention, yet that the
quality of the article is at the same time improved ;
considering also that Mr. Brunel *has obtained a patent,*
giving him the exclusive right of affording the advantages of his invention on his own terms, which, although
it gives him no power of preventing the use of it for
His Majesty's service, yet leaves him good grounds for
claiming from Government what may be deemed a
reasonable compensation for the use of his invention,
I take for granted that their Lordships have no doubt
respecting the expediency of allowing Mr. Brunel some
compensation ; and, therefore, that it is respecting only
the most eligible mode of remuneration that their
Lordships have been pleased to require my opinion.
On this supposition, therefore, it seems incumbent on
me to endeavour to devise such a mode, as should not
only prove satisfactory on the present occasion, but
which should also be calculated to afford encouragement to persons of ability in general for the production
of other inventions tending to the diminution of dockyard expenses ; while, at the same time, such remuneration should not hold up a precedent whereon
claims for compensation could be founded in any case

where the reality of the advantages had not been previously ascertained.

"In consequence of these considerations, and in conformity to the above-mentioned objects, which seem requisite to be held in view, I am induced to propose as follows:

"1st. That the time for giving the compensation should be deferred until the savings on which the claims for compensation depend have actually been realised sufficiently to ascertain their extent.

"2nd. That the amount of the compensation be made equal to the amount (as near as can be estimated) of the savings which the public will derive from the use of the invention during some specific period, such as their Lordships may be pleased to allow, and which I would venture to propose should not be less than one year.

"3rd. That for the purpose of ascertaining the amount of the saving, the average number of blocks of each description which have been actually supplied every year by contract during the last five years, be considered as the average demand for blocks for one year.

"4th. That the cost of this average number, according to the prices of the last contract, be taken as one year's expenditure for blocks, according to the present mode of obtaining them.

"5th. That as soon as Mr. Brunel's apparatus shall be reported by him to be in proper order, and the men who are to work it sufficiently trained to their work to afford a fair specimen of the despatch, and thereby of the rate of expense at which the manufacture may be continued, the proper officers on the spot be directed to note accurately the whole of the expense which shall be found to attend the making, by means of this new apparatus, a certain number of the blocks of each description.

" 6th. That in making out this expense there be noted, not only the current expense of men's labour and cost of materials, as well as of fuel for the steam engines, but also an addition of ten per cent. per annum on all the capital laid out in the machinery for giving it motion ; as also, in the way of rent, a like per centage on the cost of so much of the building in which this manufacture is carried on as is occupied for that branch of service.

" 7th. That according to the rate of expense so ascertained, there be calculated the total expense of making a sufficient number of blocks of the several descriptions requisite to supply the average yearly demand as above specified.

" 8th. That this total of expense be considered as one year's expenditure for blocks manufactured according to Mr. Brunel's method.

" 9th. That the difference between this yearly expenditure according to Mr. Brunel's mode, and that according to the present mode, estimated as above-mentioned, be considered as the yearly rate of saving which will arise from the adoption of this new mode.

" 10th. That although it appears advisable that the compensation should not be given till experience shall have afforded sufficient data for the true estimation of it ; yet, should this mode of compensation meet with their Lordships' approbation, it seems expedient that Mr. Brunel should forthwith be informed of the conditions of it, in order that he, being thereby assured that the amount of his compensation will depend entirely on the clear amount of the advantages which shall be derived from his invention, and that the time of his receiving that compensation will be no longer delayed than is necessary for the ascertaining that amount, he may have the strongest inducement to use his utmost endeavours for the completion of every part of his apparatus with the greatest despatch, as well as economy ; *whereas were the conditions of the compensation left altogether undecided*, he might, under the apprehension of not reaping the fruits of his labour, be led to direct his attention, in preference, to some other object. In favour of such a mode of compensation I would take the liberty of observing, that the greater the sum to which it may be found eventually to amount, the greater, in the same proportion, will be the advantage which the service will derive from the invention ; and the expense which such a compensation would occasion to the public would be no *new* expense, but only the continuation for a short and limited time of the same rate of expense which has heretofore been looked upon as necessary, and which, unless some such invention as this of Mr. Brunel's were to be introduced, must of course have been continued for a long and unlimited time, without any prospect of its diminution. It should also be observed, that the superior uniformity and accuracy of workmanship which would be given to the blocks manufactured by this apparatus is an advantage which would be obtained immediately, as well as continued afterwards, in addition to that of the saving of expense as above mentioned.

" In case their Lordships should think proper to adopt this mode of compensating Mr. Brunel for his invention, in giving their directions to that effect they would have only to decide on the period during which

they may be pleased to allow him the amount of the
saving according to the yearly rate estimated as above.

" With regard to Mr. Brunel's application for some
allowance for his time and travelling expenses since
he has been engaged in the erection of his apparatus ;
considering that, according to the mode of compensation which I have ventured to recommend, the
amount of it cannot be ascertained for several months
to come ; and that Mr. Brunel, in consequence of
the satisfaction afforded to their Lordships by the inspection of his apparatus in miniature, has been employed since that time in directing the execution of
this machinery here in town, as also in the superintending the erection of it at Portsmouth, I would
recommend that he should be allowed at the rate of
a guinea a day during the time he has actually been
so employed ; as also travelling expenses at the rate
of ten shillings a day, and coach hire for his journeys
to Portsmouth ; which expense, in estimating afterwards the compensation, will of course be considered
as capital sunk in the introduction of this invention.
And though I look upon this rate of payment as the
greatest which can be well allowed him, considering
the extent of pay and allowances given to persons
employed in and about the dock yards ; yet I cannot
but look upon it as the least which Mr. Brunel is
entitled to expect in consideration of the value which
would be set upon the time of a person of his talents
when employed in a private concern.

" As to the period at which the allowance should
take place, Mr. Brunel in his letter has mentioned the
latter end of August ; but as, on questioning him on
the subject, I find he has, since that period employed
a portion of his time, amounting to about three weeks,

on another business of his own, I would purpose that the 16th September should be the period from which the allowance in question should commence.

"I am Sir, your very obedient servant,
"SAMUEL BENTHAM.
"To Sir Evan Nepean, Bt.

"Mr. Brunel's letter is herewith returned."

In accordance with the suggestions made by General Bentham, the Admiralty issued the following instructions to the Navy Board.

"Admiralty Office, 7th May, 1803.
"Gentlemen,
"Whereas upon our referring to Brigadier-General Bentham an application we received from Mr. Brunel, soliciting some remuneration for the labour and expense which he had been at in the invention and perfecting of his machinery for the manufacture of blocks for the use of his Majesty's Navy, the General has, in his letter to our secretary of the 30th of last month, represented, for our information, that having examined the several articles of the machinery in question, and having seen them at work, he is fully satisfied that they are adequate to the making of blocks more perfect in regard to accuracy and uniformity of shape, as well as at a much cheaper rate, than could be made by any other means hitherto in use, and he has in consequence suggested such means for ascertaing the compensation to which Mr. Brunel might be entitled for the ingenuity and utility of his invention as appeared to him the most eligible and which can only be fairly determined by a year's trial of the machinery according to the experiments and calcula-

tions of labour, materials, and other charges pointed out by the General.

" That with regard to Mr. Brunel's application for some allowance for his time and travelling expenses since he has been engaged in the erection of the apparatus, the General has proposed, for the reasons he has set forth in his letter, that he should be allowed one guinea a day during the time he has actually been so employed; and also travelling expenses, at the rate of ten shillings a day, and coach hire for his journeys to Portsmouth; and the General has further proposed, that the 16th of September last should be the period from which the allowance in question may commence.

" We send you herewith the General's aforementioned letter, and do hereby desire and direct you (without admitting the principle recommended by the General for remunerating Mr. Brunel) to instruct the officers of the Portsmouth Dockyard, to keep account conformably to the suggestion contained in General Bentham's letter ; and you are to report to us the particulars thereof as accurately as possible, in order that by the advantages derived to the public from this invention, we may be enabled to form a judgment of the extent of the reward which may hereafter be proper to be given to Mr. Brunel for his ingenuity. And you are to make him an allowance of one guinea per diem during the time he may be employed in erecting his apparatus, to commence from the 16th of September last, agreeably to Brigadier-General Bentham's proposal ; and also an allowance of ten shillings per diem for extra expenses during his absence from town on the public service ; together with the amount of coach-hire actually incurred on his journeys, to and from

Portsmouth, upon his producing certificates from the General, of the several journeys he may undertake in completing his work.

" We are your affectionate friends,

" T. TROWBRIDGE.

" J. ADAMS.

(Copy.) " T. MARKHAM."

And at the same time the following communication was made to Brunel.

" Admiralty Office, 5th May, 1803.

" Sir,

" I am commanded by my Lords Commissioners of the Admiralty, to acquaint you, that they have given orders that an allowance be made to you of one guinea per diem during the time you may be employed in erecting your machine for the manufacture of blocks in his Majesty's dockyard at Portsmouth, to commence from the 16th of September last, and to continue until the same shall be completed ; together with the amount of coach hire actually paid, or to be paid, by you in your journeys to and from that place ; and also, an allowance of ten shillings per diem for extra expenses, during your absence from town ; and that their Lordships will consider what farther reward may be proper to be made to you for your invention, whenever the extent of the advantages likely to be derived by the public from it, shall be fully ascertained.

" I am, Sir, your very humble servant,

(Copy.) " EVAN NEPEAN."

CHAPTER VII.

1802–1810.

DIFFICULTIES IN FINDING WORKMEN, 1802–1 05 — GENERAL
BENTHAM SENT TO RUSSIA, 1805 — QUANTITY OF TIMBER RE-
QUIRED TO CONSTRUCT A SEVENTY-FOUR, 1806 — IMPEDIMENTS
TO THE OPERATIONS OF THE MACHINERY, 1807 — MACHINERY
CAPABLE OF SUPPLYING ALL THE BLOCKS FOR THE BRITISH
NAVY — REMUNERATION POSTPONED — NERVOUS FEVER, 1808 —
AMOUNT OF REMUNERATION DETERMINED, 1810 — LETTER FROM
LORD ST. VINCENT 1810 — PERFORMANCE OF THE BLOCK MA-
CHINERY. — PRESENT CONDITION OF THE MACHINERY, MARCH 1861.

THE recommendation of Sir Samuel Bentham having
been accepted as the principle on which remuneration was to be made, Brunel proceeded to perfect his
projections, and Bentham was authorised by the
Admiralty to permit him to procure "the whole of
the machinery," which was "to be paid for in conformity to the General's proposals, who is to be required to certify."

It would appear to have been now not less to the
advantage of the Government than to Brunel that
every facility should be afforded to the full development of the accepted plans. On the contrary, the
most vexatious impediments continued to be placed
in the way; and as Brunel's remuneration was to
depend upon the amount of work which the machinery
could accomplish in the course of one year, it was
matter of the utmost importance that the test should

be applied without any unnecessary delay. I shall, then, endeavour to follow the progress of Brunel's labours, that my readers may be in a condition to understand the nature of the difficulties with which he had to contend in establishing the amount of his claim to remuneration, and the grounds upon which that amount was so unreasonably modified.

The first impediment was the want of competent mechanics to construct the machinery; for this, no blame could attach anywhere, since it arose from the backward condition of the mechanical arts in the country.

In a letter to the Navy Board, September 20th, 1802, he says, "the difficulty I have met with in procuring a sufficient number of able workmen to execute the block machinery has occasioned my delaying to answer the order I have received from you." And to General Bentham, he says, "in order to execute with expedition the apparatus I am ordered to erect, I shall be under the necessity of dividing the several parts composing it to various persons. If you have any particularly able person whom I can trust with some part of the machinery, I will be much obliged to you to inform me of it.

"I take the liberty of pointing out Mr. Maudslay as a man whose abilities can be relied upon for the execution of such part of the machinery that will require the greatest exactness."

Notwithstanding the impediments presented by the want of competent workmen, Brunel, by the end of October, 1802, had been enabled to execute so much of the work as to allow him to solicit an advance of 300l. from the Navy Board. He concludes his letter thus:

"I hope that the Honourable Board will take into consideration, that I have bestowed considerable labour and expended a large sum of money in *inventing* and *completing* my first block machinery, and in obtaining the patent for it."

On the 18th March, 1803, he begs to inform the Board that he had that day forwarded the most considerable part of the machinery "by the waggon of Clark," and requests a further advance of 300*l*. And on the 19th May, 1803, having "completed the parts composing the apparatus for making blocks from seven to ten inches inclusively," he requested a further payment of 150*l*., being, he says, "part of the sum remaining due, the balance of which I shall make application for when submitting before the Honourable Board the accounts of the extra expenses I have incurred by adding several parts to the apparatus specified in the drawings which I have had the honour to lay before the Lords of his Majesty's Admiralty Board."

But now, although the machines were completed, competent hands could not be found to work them, and the Navy Board seemed incapable of seeing the necessity of procuring them. In September, 1803, Brunel, in writing to General Bentham, says:

"In consequence of the letter I have had the honour to receive from you, I have applied to Mr. Diddams for *four* block makers to assist in working the block machines. Mr. Diddams having referred my application to the Navy Board, I have been informed that *I* must hire such men.

"Being entirely unacquainted with those people, and fearful of their not being continued in that business, I cannot promise any encouragement such as to induce good workmen to leave their work."

In reply to an official application, October 7th, 1803, for an estimate of the number of blocks his machines could supply in the course of the four ensuing weeks, Brunel says, " that from the want of *four* block-makers, it would be impossible for him to form an estimate," and adds, " I cannot, therefore, answer but for what can be done by *six labourers* and *two house-carpenters*, who, if at work without interruption, at the rate of nine hours and a half per day, will be able to make, in six days, or a week, one thousand shells, and turn as many old shivers and pins.

" I beg also to observe that such labourers, who have the management of some engines, and who are attached to some part of the work, being subject to *the watch*, occasion a loss of time, prejudicial when the machines are at work."

By May, 1804, Brunel had obtained the assistance of *two* block-makers; his application to the Navy Board for a *third* " received no answer." " I have been under the necessity," he writes, May 20th, 1804, " of doing with the two men now employed; but they cannot possibly complete the work produced daily by the machines, having now another set; which, together with the first, are capable of making blocks from four to ten inches inclusive, of which size the number issued yearly, in time of war, is about 70,000." And that number, Brunel, writing to the Navy Board in March, declared his machines were competent to furnish by the 1st August, " in case all the materials should be provided."

In July, 1804, he again writes : " Though the engines for making blocks of large size will, I expect, be ready to work by the beginning of the next month, yet the assistance of block-makers is absolutely indispensable."

G

In the same communication, July, 1804, he complains " that the means for procuring brass coaks and shivers are not provided for, as they cannot be made by the present founder of the yard ; " and in August he announced to General Bentham that " without any authority from the Navy Board, he was compelled to procure them from Maudslay."

In September no steps had been taken to supply a competent founder to cast the metal coaks ; and in October 6th, he again writes to the Navy Board :

" The difficulties I have met with before I could obtain coaks cast with sufficient nicety, so as to fit without the least trimming, induces me to believe that it will be almost impracticable to be supplied with them by any contractor."

And again in November :

" I humbly request the Honourable Board to consider that the service will suffer materially, unless some means are adopted for supplying immediately the block manufactory with coaks."

In despite of the difficulties which were constantly presented to the completion of the manufactory at Portsmouth, Brunel was so satisfied that his machinery was capable of fulfilling the purpose for which it was designed, that he addressed the following letter to General Bentham, January 7th, 1805 : —

" I have the honour to submit to your consideration the project of an establishment which I think sufficient to carry on the block-making business, and capable of supplying the whole of his Majesty's navy, according to the information I have collected from accounts of the several articles issued from the six dockyards during the years 1797—1801.

" The block-making has hitherto been managed so

very badly, that it is not possible to ascertain exactly the price of every part, so as to determine the extent of the piecework of the labourers.

" This bad management is entirely owing to the want of a steam engine keeper conversant with its parts, and capable of foreseeing accidents and guarding against them.

" The present keepers are stopped by the least difficulty, and cannot point out the cause. The person who gives orders for pumping is the manager of the steam engine. Owing to his absolute ignorance, particularly in the management of the steam engine, the keepers are not able to point out any defects, and keep on till it stops entirely."

Again, on January 22nd, he writes to General Bentham in allusion to the coak-maker.

" The difficulties and delays which are experienced to procure anything wanted, make him lose as great a part of his time in applying for, as in getting them."

At length, after ineffectual efforts to obtain a sufficient number of men to work the various machines, Brunel waited on General Bentham, and represented to him the injurious effect of certain arrangements introduced into the dockyard. " I represented to the General," writes Brunel, in his journal of May 19th 1805, " that the new regulations were incompatible with the authority necessary to me to accomplish the work, and to enable me to derive all the advantage which I was entitled to for my invention.

" I had represented the same several times before, but to no effect.

" I thought of giving up the direction of the establishment unless I was furnished with proper power. I continued, however, until I had put the large engines to work, in order not to distress the service.

" After my representations to the General, he told me that I had the same power as before, yet he would not supply me with any written instructions."

Brunel then naturally expresses his disappointment that the General, " after having given his approbation to all what was done by me, was introducing regulations without my knowing the least word, and vesting all power in Mr. Burr ; which circumstance awakened a strong suspicion that I must either give up every idea of reward, or submit to terms extremely injurious to my interest and credit."

On May 25th, Brunel again sought, but in vain, to obtain from General Bentham official power to enable him to conduct the operations at Portsmouth to a successful issue, by which alone a just estimate could be formed of the saving effected to the country, and of the amount of remuneration to which he would become entitled.   He therefore addressed to the General the following letter.

" Sir,   " May 25th, 1805."

" Whereas, in the course of conversation with you on the subject of the instructions and regulations you have furnished Mr. Burr with, you have expressed your determination to have the said regulations put in force before I have completed the establishment erected by me, and also before the year's trial for ascertaining the remuneration is elapsed.

" I beg to represent to you that, as my credit and interest are intimately concerned in the best management of the materials, machinery and men, I cannot take upon myself any responsibility and any farther direction of the block-making machinery, unless I am empowered, by instructions and regulations you will furnish me with, to act as the chief of the establishment, intrusted with the management of the materials, machinery and men, until the machinery has been reported by me in proper order, and the men who are to work at it sufficiently trained to their work to afford a fair specimen of the despatch thereby, of the rate of expense at which the manufactory may be continued, and also, until the time fixed by the Right Honourable the Lords Commissioners of the Admiralty for ascertaining the amount of the saving has been completed.

"I have the honour," &c.

Brunel being unable to obtain any official recognition of his position as chief of the establishment, declined at length to exercise any farther supervision. The want of his presence, however, was quickly felt. Mr. Goodrich and Mr. Rogers, two gentlemen who best appreciated the value of Brunel's inventions, and who attached the utmost importance to the manner in which the details should be carried out, waited upon Brunel to implore him to continue his attendance. Brunel writes in his journal, May 29th,

"Mr. G. represented that nothing would go on at the wood-mills without my immediate attendance, which I had refused to pay since my arrival.

"As my resolution was not to distress either the General or the service, I had no objection to resume my functions, conditionally, that the General would supply me with some instructions on his return, or that he would not leave to Mr. Burr's option to object to anything suggested or ordered by me."

The difficulties with which Brunel had to contend

may perhaps be less attributed to any changed feelings on the part of General Bentham for him, than to a desire on the part of the Navy Board to see the growing establishment at Portsmouth assume an organised appearance, and placed under " dockyard officers."

That Board had declared itself incompetent to the superintendence of works of so great magnitude, and had placed them under General Bentham's management ; it was therefore natural that the General, upon whom so heavy a responsibility was thrown, should appoint his confidential servant Mr. Burr as general manager, while Brunel, on his part, could scarcely have felt otherwise than humbled, irritated and disappointed, to find that, however valuable as a draughtsman and overseer of workmen the individual thus selected to control his movements and over-ride his orders may have been, he was still an inexperienced mechanic, and an illiterate man.   The following extract from a letter addressed by him to Brunel, Sept. 16th, 1807, at once declares his social position :

" Sir, I'll thank you if you will speak to Mr. G——, for me, having *them* partitions across my cabin, *as* I spoke to you about."

It has been asserted by the writer in the " Mechanic's Magazine," to whom we have before referred, that Brunel, from the unworthy motive of arrogating to himself inventions to which he had no title, ventured (May 30th, 1805) to forbid Mr. Burr by written note, " to answer any queries respecting the blocks, or block-making, without his having previously communicated them to him."

In Brunel's journal the following entry appears, May 30th, 1805.

" *This request was in consequence of Mr. Burr not*

being acquainted with what the engines are intended to perform."

That the exercise of this delegated control on the part of Burr was not always judicious, may be found in the description of workmen he took upon himself to employ. While Brunel urged the necessity of obtaining the services of a few block-makers, who would readily learn how to apply the engines to the best advantage, Mr. Burr engaged on the 29th May fifteen boys, without consulting Brunel; and on the 1st July Brunel records, " since the addition of a number of boys " working at extra hours, "there are not more blocks made than at the time of short days, and less assistance ; and on the 3rd July, "since these boys were entered for the wood-mill, the circular cutters of the old scoring engine were broken by the inexperience of these workmen."

A further discreditable attempt is made to depreciate the character of Brunel, and to give undue importance to the position held by Burr. " Brunel," it is stated, " being constantly on the spot, without any immediate need for personal occupation," was enabled to gratify his vanity in " calling the attention of visitors to the machines of his own arrangement ; " and while he " had thus leisure, and opportunity for obtaining credit to himself, Burr's active and onerous duties debarred him from taking the place of a showman or vindicating Bentham's claims."

That this assertion is as ungenerous as it is unjust, Brunel's journal affords sufficient proof. Had he indeed no " personal occupation," who had embarked his present means and future prospects,—his personal character and professional reputation, in operations so novel in conception and so difficult in execution, that but one mechanist in the whole country could be found competent to execute the designs his inventive genius had furnished ?

"Non habet commercium cum virtute voluptas."

The reply of Milton to his detractors may well apply to those of Brunel.   His mornings were occupied "not in sleeping, or concocting the surfeits of an irregular feast ; but up and stirring.   In winter often ere the sound of any bell awoke men to labour or devotion : in summer as oft with the bird that first rouses, or not much tardier."   No ! Brunel was no lounger ; and however much he may have desired to have his inventions known, these intrusions were felt to be a serious tax upon his time, his thoughts, his courtesy, and his pecuniary interests.

On the 29th of May 1805 he writes : — " This frequent admission of visitors is of great hinderance to the men at work " — and again July 1st, " The place was the whole morning crowded with visitors much to the annoyance of the service.

" On the 12th went to Kentish Town to General Bentham, and represented to him the necessity of surrounding the wood-mill with a fence to prevent intruders."

This constant intrusion, against which he protested, but which he seems not to have had the power to prevent, Mr. Burr being in command of the yard, induced him to take upon himself at length on the 27th July the responsibility of ordering " the west door of the wood-mill to be shut," and to place a man at the entrance to prevent any persons being admitted " except officers in uniform."   In a sort of despair he addresses himself to Commissioner Grey.   " The works carried on at the wood-mill are considerably impeded by the number of persons who are daily admitted into the place. No regularity can be obtained when the shops are crowded with strangers. The men cannot be overlooked, and kept in that state of activity so requisite in a manufactory; I therefore take the liberty of requesting you to adopt such measures as you may deem most effectual to remedy the evil."

In July 1805, Sir Samuel Bentham was suddenly withdrawn from his duties as inspector-general of naval works, to undertake a mission to Russia, for the purpose of superintending the building of ships of the line and frigates for the British navy; the demands of the service exceeding the capability of the Government establishments to furnish a supply.* Provided with a staff of officers, mechanists, and tools, he arrived at St. Petersburg, where, to his amazement, disappointment and disgust, he learnt that no provision had been made for his reception; and where "even our ambassador had not been apprised of his mission — diplomatic assurances of readiness to oblige having been construed into a full acceptance of the proposal of our Government to build ships of war in Russia."

* According to the report of the surveyor-general, it was calculated, that to construct a 74-gun ship, it required 2000 tons of timber, or about 3000 loads; and the average duration of a vessel was supposed to be 25 years. The oak trees, destined for the construction of ships of the line, are required to be of the full growth of 100 years. Such trees are calculated to yield a load and a half of timber each; therefore, for a seventy-four, there would be required 2000 trees. Now an acre of land can sustain about forty oaks; hence 50 acres are required to furnish the supply for one vessel. But as the quantity of waste lands in the kingdom amounted, in 1806, to about 20,000,000 of acres, there ought to have been no difficulty in securing the necessary supply.

During Sir Samuel's absence, Mr. Goodrich, his mechanist, a gentleman who, we have already seen, fully appreciated the importance of Brunel's inventions, was appointed *locum tenens* at Portsmouth, to Brunel's obvious relief; for he at once writes to Mr. Goodrich,

" Dear Sir,

" I am very happy to hear that the management of the business of the General's office, and particularly of the several works at present in hand, has been intrusted to you; and hope that as soon as the bustle which has been created by the sudden departure of the General is entirely over, we shall be able to consider upon the means requisite to complete the work under my direction."

The necessity which at first justified the use of some of the older machines could no longer be urged. Experience had shown that the alternate motion of a saw frame could not be increased beyond a certain number of strokes in a given time, without endangering the machinery. Brunel, therefore, abandoned that system, and adopted the rotatory. One circular saw of his construction was found to do all the work of twenty-four saws in six frames : elm logs 24 to 36 inches diameter were converted into various materials by a saw moving at the rate of 108 to 112 revolutions per minute.

With the progress of the war the demand for blocks increased so rapidly that no ordinary means could have furnished the supply. Skilled workmen were few, and therefore costly, and the work was often dependent on their caprice. And although Brunel's machinery was really equal to the emergency, the full results could not be obtained. The steam engine frequently failed in its duty, and when capable of working, "was constantly waiting for materials, *and the orders sent to the mill could not be taken in hand.*" Again, when the supply of timber *was* obtained, the quality often proved inapplicable to blocks.

" I have represented to the officer of the yard," writes Brunel, "that the elm was too rough, and too strong ; and that the consequence would be, that the shells of blocks made out of such materials would shrink in seasoning, and could not be depended on in service. They informed me there was no other then to be had." On the 17th November, 1806, he writes to Mr. Goodrich, " The orders for blocks have not for several weeks been complied with, for want of brass coaks of every size ; " and on May 11th, 1807, he writes, " The wood sent to me to be converted, having been used for other purposes before, *is full of nails.*" As a consequence the saw was frequently deranged, the edge reduced $\frac{1}{16}$ of an inch, and the work vexatiously delayed.

On Oct. 4th, 1807, he complains of the want of brass coaks, and the irregularity to which the work is subject, " for the men being obliged to alter frequently their engines, cannot make their wages upon piece work, and the *orders cannot be executed.*"

Also, in a letter to the Navy Board, August 18th, 1808, he says, " I beg leave to inform the honourable Board that from want of proper supplies of *lignum vitæ* of large dimensions, the mill is working to considerable disadvantage. The necessity of making some articles in preference to others, prevents the wood from being appropriated and converted according to its size and soundness ; consequently, the quality of the work produced is materially affected. On the other hand the machines, not being properly supplied with materials, are not kept regularly employed."

As the credit of the machinery was thus invalidated, so in like manner its pecuniary value was sought to be depreciated by estimating the product, not according to prices of materials when the work was undertaken, and which were paid by the old contractors, but according to what those prices had become, after five years of hard struggle, against the ambition of Napoleon, with wheat from 78 to 94 shillings a quarter, with elm timber at 100 shillings the load in place of 52s. 6d., *lignum vitæ* at 23l. in place of 5l. and 10l., brass at 1s. 1d. per lb., in place of 5d., and the wages of skilled workmen at 3l. 10s. to 4l. a-week! No wonder that the savings effected by Brunel's machines should have been variously estimated from 6,691l. to 26,000l. per annum.

To Brunel's repeated application for a settlement, the Navy Board excused itself by assigning as a reason (January, 1809), " the many additions and alterations which *Brunel* had found it necessary to make in his machinery," entirely overlooking the fact, as stated by Brunel in his reply, January 20th, 1809), " that it was in consequence of the increase in his Majesty's navy, *that it became necessary* to give such a disposition to the buildings and machinery as to enable it (the machinery) to supply a much greater proportion of work than was at first calculated," and that " the additions, alterations, and improvements were much more difficult to be carried into execution than the first erection." Indeed, it is obvious that with the progress of the war, the value of the machinery became more apparent, " directions were therefore," as Brunel states, " constantly reiterated by every administration at the Admiralty, and every other channel of communication, that I was not to fail to possess the machinery with powers commensurate with the increasing exigencies of the service, and to secure Government against the accidental obligation of having recourse to contractors for *any* article of the block manufacture, thereby enabling the Navy Board to supply the wants of the Ordnance Department, which it had been intimated to me they had in contemplation."

In point of fact, during the year 1808, the machinery actually furnished " 130,000 blocks of every size and species, and also a certain proportion of the articles comprehended in the block-maker's contract; " the value of these articles was estimated at 50,000*l.*, and in January, 1809, it was found that the quantity manufactured during the last three months of 1808 would give an average of 160,000 blocks annually, the value of which was estimated at not less than 54,000*l.* The machinery was complete, it was capable of furnishing all the blocks heretofore supplied by Portsmouth, Plymouth, Deptford, Woolwich, Chatham, and Sheerness. The officers of the yard had so reported; yet the genius which devised, and the skill which organised all this, was still to be measured by a standard which the exigencies of the time had virtually destroyed, and he who had been his adopted country's benefactor, was doomed to suffer all the misery of hope deferred, to have his claims to remuneration constantly postponed, and that recompense denied to which his originality and his talents had so largely entitled him.

The distress to which Brunel was reduced began seriously to affect his health. He was attacked with nervous fever, and for some weeks was unable to turn his attention to any business. " The duty of a father," he says in a letter to the Admiralty, " whose anxiety for the welfare of a young family prompts me to reiterate my solicitations, and at the same time to represent to their lordships the uncertain and unsettled state I am kept in is, in every respect, extremely injurious to the interests of my family, being prevented from engaging in more extensive concerns." And to Lord Mulgrave he writes, July 13th, 1808, " engaged as I have been lately, merely in the pursuit of information on the fate of my late application respecting my remuneration for my past services, I find that the best part of days and weeks are wasted away without any appearance of success ; and that being thereby prevented from paying immediate attention to the applications I have been honoured with, I may lose a favourable opportunity of reaping some advantage from my abilities."

At length, in August, 1808, the Admiralty directed the payment, not of the promised remuneration, but of *one thousand pounds on account!* Brunel was in despair. He had already expended more than double that sum upon " models, drawings, and experiments." On the 20th August, in acknowledging the receipt of this fraction of his just claim, he says, " I would humbly beg leave to represent to your lordships, that it would be an irretrievable loss to my family if I were obliged, from motives of prudence, to relinquish my present pursuits, and a loss which I should feel still more acutely, if I were to reflect, that the whole fruit of my exertions and abilities in the service of the Government, could not procure me the means of securing the advantages of a concern*, in which I have lately engaged, being, at the time I entered into it, fully persuaded that nothing could prevent or delay the settlement of a remuneration already determined upon by your lordships ; " and to Mr. W. W. Pole (Admiralty), he says, June 13th, 1808, " I beg leave

---

* Saw-mills on a large scale, erected under Brunel's direction at Chelsea, and in which he held a considerable share.

to represent that the advances made by me for models, drawings, and experiments for the various objects upon which I have been employed in the service of Government, *exceed the sums I have received,* at the rate of one guinea a day, from the 16th September, 1802, to this day!" that is, for the six years during which his time and thoughts had been almost exclusively devoted to the service of the country.

To the Messrs. Borthwick also, who had engaged him to supply designs for an extensive saw-mill at Leith, he writes, March 26th, 1810 —

"You may imagine that I have made little progress in your plans. It is with great regret I find a considerable portion of my time still taken up in the settlement of my business with Government, to the very great detriment of my private concerns."

How little do those in authority sometimes reflect on the suffering inflicted upon men of genius, of whose gifts they have availed themselves, in resisting or repudiating claims admitted to be just; and thus, through thoughtlessness, or wilfulness, bringing perhaps ruin on the victim, and certainly discredit upon the country, of which they are supposed to be the honoured guardians.

> "Full little knowest thou who hast not try'd,
> What hell it is in suing long to byde;
> To lose good days that might be better spent,
> To waste long nights in pensive discontent:
> To speed to-day, to be put back to-morrow,
> To feed on hope, to pine with fear and sorrow;
>   *   *   *   *   *
> To have thy asking yet wait many years,
> To fret thy soul with crosses, and with cares,
> To eat thy heart with comfortless despairs." *

---

* Spenser's Mother Hubbard's Tale.

At length the Admiralty consented to accept the following approximate estimate :

	£	s.	d.
Amount of all articles manufactured by the block machinery, and delivered into store between the 1st of July and 31st of December 1808, computed on the contract arrangements	24,362	12	10
Brass work, shivers, &c., at 13¼d.	1,023	15	8
	25,386	8	6
Advance of 8 per cent.	2,030	18	3
	27,417	6	9
Deduction for fees, 1½ per cent.	411	5	2
	27,006	1	7

Amount of materials used :—	£	s.	d.			
Elm, 948 loads, @ £2 12 6 . .	2,488	10	0			
Various materials	141	11	8			
Brass, 50,878 lbs., @ 13d. .	2,755	17	10			
Brass, 18,544 lbs., @ 9½d. . .	734	0	8			
Lignum vitæ, 205 t. 14 cwt. @ £21	4,319	14	0			
Iron, 40 t. 6 cwt. 3 qrs. 1 lb., @ £26	1,048	15	9			
Wages	3,418	19	8			
10 per cent. on £45,000 capital, for 6 months	2,250	0	0			
Interest on £20,000 floating capital @ 5 per cent. for 6 months .	500	0	0			
Coals for steam engine . . .	296	8	0			
Sundry expenses of steam engine	100	2	0			
Cordage, leather, bands, &c. .	462	7	1	18,516	· 6	8
Difference				8,489	14	11
Amount of offals obtained				342	0	0
Savings for six months				8,831	14	11
Savings for a year				£17,663	19	0

From this, however, nearly 600*l.* was still deducted, and ultimately, the total sum received by Brunel, August 21st, 1810, after seven years of labour, anxiety, expectation, vexation and disappointment, was 17,093*l.* 18*s.* 4*d.*, in the following manner :

				£	s.	d.
1808	.	.	.	1,000	0	0
1808	.	.	.	1,000	0	0
1809	.	.	.	5,000	0	0
1810	.	.	.	10,093	18	4
				£17,663	9	10

That those highest in authority sympathised with Brunel at the disappointment he was made to experience at the hands of the Admiralty and Naval Board, there is good reason to believe.

Lord Spencer had more than once declared that the least amount of remuneration to which he was entitled, was 20,000*l*. ; and Lord St. Vincent, writing to him on September 3rd, 1810, after his mind had been relieved by the final settlement of his claims, says —

" I am very much gratified by your letter of the 3rd, which had removed a good deal of apprehension from my mind, on account of the delay in granting to you the remuneration due to the services you have rendered. Am I to understand that you are no longer to superintend the ingenious and useful work in Portsmouth yard ? I much fear it will not prosper so well in other hands. Heartily wishing you success in all your undertakings,

" I remain very much your humble servant,

" St. Vincent.

"Racketts : September 5th, 1810."

For a description of the block machinery the reader is referred to Appendix B. We shall here only add that where FIFTY MEN were necessary to complete the shells of blocks previous to the erection of Brunel's machinery, FOUR MEN only are now required ; and that

H

to prepare the sheaves, SIX MEN can now do the work which formerly demanded the labour of SIXTY.

So that TEN MEN, by the aid of this machinery, can accomplish with uniformity, celerity, and ease, what formerly required the uncertain labour of ONE HUNDRED AND TEN.

When we call to mind that at the time these works were executed, mechanical engineering was only in its infancy, we are filled with amazement at the sagacity and skill that should have so far anticipated the progress of the age, as to leave scarcely any room during half a century, for the introduction of any improvement. Mr. W. Chisman, in a report to Admiral Grey, March 23rd, 1861, and forwarded by him to Lord Clarence Paget, who kindly handed it to Sir Benj. Hawes, K.C.B., says—" The block machinery is at present in good working order. The machinery, in every main particular, is as the inventor left it. Considered in an economical point of view, the value of the present block machinery is *incalculable*, and it will be found that, if the expense of blocks made at this yard is compared with the list of prices of Mr. Esdaile's block manufactories of London, it will not, in some sizes, amount to one half the cost." Mr. Chisman farther states, " that the superiority of these blocks over those made in the merchant service or used in foreign ships, is evident, as machinery is calculated to prevent the consequences of inaccurate workmanship when made by hand."

CHAPTER VIII.

1805–1816.

APPARATUS FOR BENDING TIMBER, 1805 — MACHINERY FOR SAW-
ING TIMBER, 1805–8 — BIRTH OF A SON, 1806 — MACHINE FOR
CUTTING STAVES, 1807 — WORKS AT WOOLWICH, 1808 — THE
SAW — SKETCH OF ITS HISTORY — SAW MILLS, BATTERSEA. —
VENEER ENGINE, 1808 — HAT AND PILL BOXES — IMPROVE-
MENTS IN MOTIVE POWER, 1810 — COMPARISON OF COST OF SAW-
ING BY MACHINERY AND BY HAND, 1811 — DESIGNS FOR CHAT-
HAM, 1812 — DIFFICULTIES — DESCRIPTION OF MACHINERY —
BACON — MR. ELLACOMBE, 1816 — PRESENT STATE OF THE WORKS,
MARCH 1861.

IN 1805 Brunel suggested an apparatus for *bending* timber. " In the mode commonly practised," he says, " the plank or scantling, after being heated for a certain time, is taken from the kiln, and then moulded. It is obvious that the fibres, strained by this operation, and receiving no further assistance from the emollient effect of the steam or water, are exposed to separate, and even to break." To obviate this, he proposed to erect an apparatus, by means of which, the timber would be completely softened by the action of steam or boiling water, and subsequently moulded by a slow process to any given form, " without being removed from the place during the operation."

In this year also he took out a patent " for sawing timber " in an easy and expeditious manner. The im-provement consisted in the modes of laying, and holding the piece of wood in the carriage or drag ; in the facility of shifting the saw from one cut to another ; and in the practicability of sawing both ways, either towards or from the saw or saws."

In 1806 Brunel patented his machine for cutting veneers or thin boards; and in 1808 he secured another patent for " Circular Saws." These saws were " intended to cut out thin boards, or slips, with as little waste as appears practicable." They were to possess the two essential qualities of a thin edge and great steadiness. In 1812 and 1813 he took out two other patents for improvements in " Saw Mills." The improvements consisted :

1. In accurately fixing the saw or saws in the frame.

2. The preventing the heads of the frame being bent.

3. The adjusting the inclination of the frame so as to relieve the friction of the saws when ascending.

4. In allowing the saws to be shifted any distance from each other according to the nature of the work required to be done.

5. In keeping the cutting edge of the saw true; and

6. In permitting all the saws in a frame to be whetted with an equal degree of precision by the application of a special apparatus.

These improvements completed his conceptions of all the requirements of a sawing machine.

In 1806 Brunel entered into negotiations for the organisation of a sawing establishment at Battersea, and for a blockmaking establishment which should be capable of supplying the whole of the commercial shipping. The former of these only was carried into effect. He had taken out a patent for a circular knife for cutting veneers or thin boards. The application of this instrument proved most successful. Boards from one inch to the one sixtieth part of an inch in thickness could be cut from the timber without leaving any waste. The slabs thus cut presented a beautiful polished appearance and required no subsequent preparation.

The memory of one event in this year was specially to be cherished. " On the 9th of April, and at five minutes before one o'clock in the morning," Brunel records in his journal, " my dear Sophia was brought to bed of a boy." His joy was however sadly dimmed by the heartless conduct of the government in withholding from him, as we have seen, his just reward, and thereby subjecting him to pecuniary difficulties and anxieties, against which his services ought to have been his protection. That which should have brought the highest happiness of which his domestic nature was capable, now only added another pang to his loving heart. Time, with healing on its wings, opened to him at length a brighter day, when his fondest hopes were realised, and when, at an age at which others have scarcely passed their novitiate, this child, Isambard Kingdom, achieved a name, and attained a position, that placed him in the first rank of his profession. It may indeed be said that by the boldness and magnitude of his works he came ultimately to be con founded with his father, and to have qualifications and successes assigned to him which he himself never claimed, and to which the father alone was preeminently entitled.

In 1807 Brunel submitted to the Victualling Board designs and estimates for machinery applicable to the cutting of staves for the use of the cooperage belonging to that department. The machinery was calculated

to cut about *sixty thousand dozen staves*, of various
sizes, for dry and light casks, during the day of *twelve*
hours, with the assistance of seven men and a saw-
whetter.

In the following year he was called upon by the Ordnance Department to furnish designs for saw-mills for
Woolwich, a call which he at once obeyed ; and in ex-
plaining his designs he says, " they are the result of experience, and not merely copies of what I have already
executed ; but are mostly original ideas peculiarly
calculated to answer for the service of the gun carriage
department, and adapted to the local disposition of the
ground." He farther adds, "although I cannot claim the
merit of original invention in the saw-mills, I would beg
leave to observe, that saw-mills, such as those used on
the Continent and in America, are confined to a uniform work, and entirely to fir. They could not be
used with any advantage in the service of the carriage
department for which elm, ash and oak timber,
varying in size, length and form, are indiscriminately
used, and to be converted into scantling of different
dimensions." *

Brunel's large saws were capable of cutting at the

---

* The saw is of high antiquity. The inventor was honoured
with a place amongst the gods in the mythology of the Greeks.
According to Beckmann, saw mills were in operation at Augsburg
in 1322, and in Norway about 1530, where they were called the
" New Art of Manufacturing Timber." When a saw mill was introduced, about 1663, into England, by a Dutchman, it encountered an
opposition as determined as that given to printing in Turkey, to the
ribbon loom in the Roman States, and to the crane in Strasburg. A
mill erected near London was abandoned in consequence of a combination of sawyers. Subsequently, in 1767–8, Mr. John Houghton,
a wealthy merchant, erected a mill at Limehouse ; but a mob assembled and utterly destroyed the building and machines.

rate of ten to twelve feet per minute, with the attendance of one person only. As a consequence, the price of sawing straight timber was at once reduced from 3s. per hundred to 6d., or one sixth ; and by applying the machine used at Portsmouth for making iron pins for blocks to the formation of the axles of gun carriages, which had heretofore been made by hand, he reduced the cost from 3s. to 3d.; or one twelfth. This was a more brilliant result than that which has attended all the improvements in another and justly extolled branch of industry, the making of pins *, and was so recognised by those in authority. For the valuable services which Brunel rendered to the Royal Arsenal at Woolwich he received a grant of 4500l.

In the spring of this year Brunel removed to Lindsey Row, Chelsea. At Battersea, on the opposite bank of the river, he had erected saw-mills in which he had invested all his available money, with the reasonable hope that he had secured, by so doing, a valuable provision for his increasing family and his own declining years ; a hope, however, never to be realised.

It was at these mills that the process of rabbeting or grooving timber with profit, was first accomplished. By the common mode, the rabbet is cut out in chips, which were worthless. By means of the circular saw, the slip of wood, the removal of which forms the groove, is at once separated in its entirety, so as to be available for laths, rails, &c., the value of which exceeded the expense of the operation. Here also the veneer and conical engines, or plain and conical knives, were first tried and perfected. Slabs of hard wood, as mahogany for tables, and shavings for hat and pill boxes, were cut from the

---

* Babbage, Economy of Machinery.

solid timber, forty and fifty to the inch, As a consequence, wood began to be substituted for paper in the manufacture of those articles.

To these important economic advantages to the public was added the high gratification to Brunel himself of being able to employ children in the manufactory. The love of the young was a distinguishing and abiding feature in Brunel's character ; and now, after a few hours' instruction and one day's practice, he had the happiness to reflect, that for a large number of these special objects of his sympathy he had provided the means of earning for themselves an honest and sufficient livelihood.

Four very important results may then be attributed to the application of these engines.

1. The reduction, at once, of the price of furniture made from costly woods.

2. The diminution of the consumption of paper.

3. The advantageous introduction of the labour of children in making boxes, &c.

4. The retention within the kingdom of the money heretofore paid for those articles in Holland, Germany, and the Tyrol, from whence they had for years been largely imported.

In 1810 Brunel took out a patent for " Improvements in obtaining Motive Power." This was proposed to be effected by employing the inclined hollow screw (the screw of Archimedes) to force atmospheric air into a vessel of *cold* water, from which it was to escape into an inverted funnel, and thence through a pipe to be conveyed to another vessel containing *hot* water. In this vessel a bucket-wheel was to revolve ; the air conducted through the pipe, and rarefied in its passage through the heated water, to ascend beneath the buckets, and by its buoyancy to give motion to the wheel, as water operates upon an overshot wheel in the open air. In the words of the patent, "the ascending air, by being brought into contact with the hot water doth become greatly enlarged in its dimensions beyond the dimensions of the same air when forced down through the cold water, and that, in consequence thereof, the mechanical force communicated to the wheel by the ascending air is much greater than the force required to turn or give motion to the Archimedes screw, instrument or organ employed in forcing the same down when cold ; and that I do employ a part of the said greater force in giving motion to the said screw, instrument or organ, and the remaining part in giving motion to machinery."

It does not appear that any practical result was obtained from this invention.

In November 1811, Brunel submitted a comparative statement to the Navy Board, which exhibited in a strong light the importance of substituting machinery in our dockyards for the handwork then in use. In sawing wood 600 men were employed, or 300 pairs of sawyers. The wages paid were, to top men 3s. 6d. per day, pitmen 2s. 10½d. = 6s. 4½d. But, as these men usually worked by the piece, they were enabled to realise 10s. per day, which would represent for each pair of sawyers 220 feet of sawed material ; and 220 × 300 would give 66,000 feet as the total quantity produced daily, something more than three fifths of which, or 40,000 feet, was assumed as fit to be cut by a saw mill.

It was found that a saw mill composed of eight saw frames and carrying an average of thirty-six saws, would produce twenty-one superficial feet of sawing per minute,

or 1260 feet per hour. Four mills, therefore, would
supply all the requirements of Portsmouth, Plymouth,
Chatham, Woolwich, Sheerness, and Deptford. The
average cost of one such mill was computed at 11,400*l.*

Interest on capital, and wear and tear .	.	10 per cent.
Expenses of working, saws, files, &c. .	.	13¼ ,, ,,
		23¼

Or say 2650*l.*

The average rate paid for sawing was 4*s.* 2*d.* per
hundred feet. Say for 10,000 feet 20*l.* 16*s.* 8*d.*
per day, and for 300 days 6,250*l.* per annum. De-
ducting the cost of the same amount of millwork, we
have 6,250*l.*, less 2,650*l.*, equal to 3,600*l.*; which, for
four mills, would give a total saving in this manufacture alone of 14,400*l.* per annum!

The success which attended Brunel's improvements
in the construction of wood mills aroused not only the
indignation of the sawyers—a large and influential class
in those days—but the jealous passions of some of his
contemporaries. Government was condemned for the
encouragement it was affording to a foreigner whose
mills were pronounced to be *expensive, strapbreaking,
crankbreaking* machines, and an unfavourable comparison was instituted between them and the cheap and
efficient Scotch mills. The erection of one of these
Scotch mills by Mr. Rennie, one of the first civil engineers of the day, at Rotherhithe, soon permitted the
government to ascertain their real value. This mill
had cost about 9,300*l.* : the building is described as
"indifferent." A ten-horse steam engine worked four
frames, and the week's produce was reported to be
about 9500 feet of sawing, something less than the

work of five pairs of sawyers : while the smallest frame in the saw mill at Woolwich, a mill with six frames, employed upon timber similar to that supplied at the Rotherhithe mill, was cutting from 10,000 to 12,000 feet in the week. The building was of a neat, substantial character, the machinery of the first class, the power 20 horses, and the cost 12,000l. The unequivocal evidence which Brunel had afforded of his superiority as a mechanical engineer, again induced the government to secure his services in carrying out proposed improvements in the dock yard at Chatham. After various negotiations, an order was given to him, on the 31st January 1812, to provide the necessary machinery and steam engine.

But it was not sufficient that his machinery was sanctioned ; he desired to adapt it, in the most economical manner, to the locality in which it was to be erected. Accordingly he visited Chatham with this view. Alterations and improvements had indeed been suggested there by some of the most celebrated engineers ; but they all contemplated an increase of space. When the plan last submitted was explained to Brunel by one of the officers, he observed, " Ah ! then, gentlemen, you would require the seven-league boots of the Marquis of Carabas to go from one end of the yard to the other."* After a careful examination of the ground, he suddenly exclaimed, " That hill must be bought ! " " Oh yes," replied the officer he had before addressed, " government has been so recommended ; but the cost of the

---

* " Lorsqu'il faut aller," says Baron Dupin (*Force Navale de la Grande Bretagne*), "plusieurs fois par jour, d'une extrémité de ces vastes établissements, à l'extrémité opposée, l'on perd un temps infini ; l'on se trouve épuisé pour s'être rendu seulement sur les lieux où il importe de déployer la plus grande, la plus constante activité."

removal of so great a mass of earth has deterred them."
' Remove ! take away that noble hill — the most valuable bit of ground in all the yard !" replied Brunel : "No,
no ! but buy it — buy it as quickly as possible." Little
could those who witnessed this short dialogue, recounted
to me by Brunel himself, with all the animation belonging to the expression of a new idea — little could they
have predicted the important part which " *that hill,*"
would play in the hands of Brunel, or the transformation
it was destined to undergo at the touch of the magician's
wand.   We will pause for a moment to cast a retrospective glance at the service of this yard at the time
Brunel was called upon for his designs.

The quantity of foreign timber required to be converted annually in Chatham yard was about 8000
loads.   The logs were landed at the slips, and dragged
to some convenient place for survey, where they remained till they were again dragged to any place where
room for stacking them could be found.   " For the
landing of these 8000 loads of timber," says Brunel, in
his report to the Admiralty, " there is required at least
6000 goings and comings of teams of horses, merely to
lay the timber for survey — 6000 times to and from the
stacks — at least as many more times one hundred
yards in aiding the lifting on the stacks."   From the
stacks the timber had to be transferred to the saw mill,
and again from the saw mill, when converted, to be
stacked for use.   " So that the expense attending the
land carriage of 8000 loads of timber could not be estimated at less than 10*s*. per load," and for the 8000
loads 4000*l* !

Brunel proposed first to extend the use of a steam
engine, to be erected to work his machinery, to the
landing of foreign timber.

2. To the carrying of the same and laying it for survey.

3. To the stacking of it.

4. To the taking it from the stack and conveying it to the mill; and

5. To the removing of the scantling (wood when cut), after conversion to a convenient distance from the mill, so as to prevent any obstruction near it.

Through "*that hill*," about 2000 feet in length and 200 in breadth, and which rose 38 feet immediately above the Medway at high water, Brunel proposed to sink an elliptical shaft, the diameters to be 90 feet, and 72 feet, to the level of the bed of the river, and from the bottom of this shaft to run a subterraneous canal, or tunnel through the mast pond to the river; the bottom of the shaft therefore to become a reservoir into which the timber could be floated from the river, freed from the sand and gravel which it so plentifully collected in the dragging process through the yard, and which so largely impeded the operation of the saw. Round the reservoir was to be a passage on a line with the water. Above the reservoir he proposed to erect an apparatus having a communication with the top by steps, for elevating the logs to the highest point of the yard. At a short distance from thence the saw mill was to be placed. From one side of the shaft and parallel to the boundary wall of the yard, *an iron railway*, supported on a double row of iron pillars, was to proceed, along the edge of which, beds were to be formed to receive the timber to be surveyed. It is worthy of remark that we have here a railway, the construction of which anticipated by ten years the enterprising project of Mr. Joseph Sanders at Liverpool, and which, curiously enough, was laid down on longitudinal timbers in the manner subsequently adopted for the Great Western line. Below the saw mill were to proceed other railways, on a gentle incline, to convey the timber, when cut, to the lower part of the yard, where it was to be stacked for use. A steam engine of 32-horse power was to be the moving force applied to these various operations.

In August 1812, Brunel's terms were accepted, and the work, as above detailed, was at once proceeded with. Much delay arose, however, from the want of skill and attention on the part of those to whom the execution of the tunnel, the reservoir, and great chimney of the steam engine were intrusted by the authorities; those works being taken out of Brunel's hands. In the tunnel, his design had been altered without his sanction, and his instructions neglected. An elliptical arch was substituted for a segmental, and the sides or abutments were built vertical, when they should have battered or inclined. The brick work also had been shamefully put together. It was no wonder, therefore, that it gave way. For a distance of 45 feet the arch fell, killing one man, and injuring ten. The chimney, 120 feet in height, having been founded partly on clay and partly on chalk, settled irregularly, and continued to be to Brunel a source of anxiety for months, during which it underwent a variety of doctoring to sustain its congenital weakness.

On the 27th October 1813, Brunel writes to Mr. Holl (Navy Board): " Mr. Vinall (Master Bricklayer at Chatham), expresses his confidence in its safety. I have nevertheless taken upon me to stop the progress until you decide on it." Brunel also suggested "that a kind of ladder be formed inside the chimney, with wrought iron bars let into the angles, at 18 or 20 inches apart, in order to be able to reach the top of the chimney at any time when required."

On the 2nd November, 1813, a report, signed by G. Parkin and P. Hellyer, and which had the concurrence of Sir Robert Seppings (who more than once took upon him to ridicule Brunel for his chimney fears), was addressed to the Commissioners of the Navy. " We beg leave to state that the chimney is, in our opinion, in a state of perfect safety; that it may be proceeded with without fear. In referring to Mr. Brunel's remarks, we have to observe that the chimney stands on a firm foundation," &c. As a consequence of this report, Sir Robert Barlow replied, that " from what is therein set forth, added to my own observation on the actual state of the building, I should recommend proceeding to the completion of it (the chimney), agreeably to the original design." ! !

The work was accordingly proceeded with. On the 18th of February following, Brunel records in his journal: " Observed that the chimney cracks very much, and continues to bulge out. Mr. Vinall is *now* sensible of it, and proposes to prop it by means of buttresses; and I propose, in addition, *some wrought-iron ties*, which may be buried into the brick work." Not only was the chimney thus secured, and Brunel's observation and judgment vindicated, but a principle of building was suggested, which did not receive its full development until twenty years afterwards, when the experimental arch, to be hereafter described, was constructed.

Fortunately for Brunel, the enormous cast-iron tanks, as provision against fire, erected over each wing of the saw mill, met with more consideration, or they might have shared the same fate as those at Deptford, which, from defective support, gave way.

The faulty design of those tanks was often condemned by Brunel. The sides were upright, the water six feet deep. "Yes," he would say, "a good long winter's frost, and they will bring their own destruction." The tanks at Chatham were very different; the sides inclined at an angle of 30°, thus permitting the expansion of their contents without injury.

Amidst these delays and discussions, upon matters in no way under Brunel's control, the hydraulic apparatus for raising the timber from the reservoir had been erected. From the bottom of the basin rose two cast-iron standards, between which, on the same shaft or axle, were suspended what was termed an *elevator* and its *counterpoise*, by means of chains passing over pulleys at a height of sixty feet, and coiled round drum-wheels. The *elevator* consisted of a sort of platform, upon which the balks of timber were floated.

The *counterpoise* was formed by a tank, capable of containing ten tons of water, and which worked into a small reservoir, divided from the larger one by a dwarf wall. The drums of the elevator were considerably larger than those of the counterpoise, and the chains stronger and heavier. When a balk had to be raised, the elevator was allowed to descend, raising at the same time the empty counterpoise tank, up to a point where it was supplied with the waste water from the steam-engine; when the balk was brought upon the elevator, water was admitted to the counterpoise tank sufficient to overcome the weight of the balk to be raised; the tank then descended and the balk rose. This duty accomplished, the tank discharged its water into its own reservoir, to be again elevated in its turn. To effect this simple arrangement with precision and economy, was not however an easy task. The supply of fresh water was small, and to prevent any loss the following arrangement was made.

Subterraneous tanks were formed to receive the waste water and steam from the steam-engine, in which they were condensed and cooled, and from which the water was conveyed to the counterpoise tanks. Having performed its function in the manner described, it was returned by a siphon, seven inches diameter, to the well of the steam-engine, to be once more converted into steam, and again conveyed to the counterpoise to complete a continuous circle of economic duty.

The regulation of the machinery, simple as it appeared, called for a further exercise of ingenuity. In the first operations, the chains of the counterpoise increasing in length and weight as they descended, and those of the elevator decreasing still more quickly as they ascended, caused the motion to be so much accelerated as to overpower the break, and allow the timber to fly up with a fearful velocity. This was however speedily remedied by Mr. *Ellicombe* (so the name was then spelt), the accomplished resident engineer, of whom more hereafter. He suspended one end of a chain to the bottom of the timber carriage or elevator, equal in weight to the two chains of the counterpoise, and the other end to a hook about half-way up one of the standards which carried the drums ; this had the effect of checking the acceleration, and of securing a uniform motion. A more elegant contrivance was subsequently employed by Brunel : this was an apparatus similar to the governor of a steam-engine, which, being made to revolve in water, rendered the whole action of the machine as smooth as the motion of oiled surfaces upon one another.

I

The largest as well as the smallest logs of timber were raised from the reservoir in precisely the same time, *thirty seconds*. Indeed, a balk sixty feet long and sixteen inches square has been raised in *twenty seconds*. Arrived at the top, the huge arm of a gigantic moveable crane, of a peculiar construction, seized the log, descended the inclined railway by its own gravity, and gently deposited its burthen on the drying-beds to be surveyed; being drawn back by a chain passing round a drum worked by the steam-engine, if necessary with a rejected log, which it lowered to the reservoir by the same means with which it had been elevated, to be expelled to the mast pond and to the river.

One man sufficed to direct all the operations of this enormous machine ; to move it, stop it, and turn its arms or jibs in any required direction. The same crane was also employed to convey the dried timber to the mill. This mill consisted of eight saw frames and two circular saw benches, with the necessary windlasses and capstans, and was capable of accomplishing the work of fifty pairs of sawyers. When converted, the scantling or sawn timber was conveyed to any part of the yard on trucks, and piled for use, where it was so arranged along each side of inclined planes, that it might be examined with perfect ease, the sizes selected, and conveyed to the docks or slips by single horse trucks, without in the slightest degree interfering with the other work in the yard. From the moment the timber was raised from the water, not a single piece required to be even turned, until placed on the trucks, to be forwarded in its manufactured state to its destination : ten to twelve men, with the aid of this beautiful machinery, performing all the operations once so costly, and with a regularity, precision and celerity never before attained in works of such character and magnitude.

We can well understand the importance which Brunel attached to the supervision of this novel and comprehensive system of machinery, and his anxious solicitude that it should be placed in full and efficient operation, before it was handed over to dockyard officials, to whom its mechanical arrangements were but an innovation, and by whom its value was only partially understood. Certain it is, that long before the works were reported by the inventor as complete, an effort was made to deprive him of the services, first of Mr. Bacon, and subsequently, with more success, of Mr. Ellicombe, two of his most able and confidential assistants. When his application to have Bacon appointed to the charge of the saw mill was rejected, he was filled with anxiety and disappointment, dreading a repetition of what he had already experienced during the erection of the block machinery at Portsmouth. It was Bacon who appears to have been first sent to Scotland, in 1811, to superintend the saw mills erected for the Messrs. Borthwick, from whom he received the highest character. It was through them he was subsequently transferred to Woolwich Arsenal and ultimately, in 1813, to Chatham; and on Bacon therefore he could implicitly rely, not only for strict integrity and for a large amount of mechanical skill, but for full and accurate appreciation of the relation subsisting between the several parts of a well considered system.

"It is frequently so long," Brunel writes to the Navy Board (October 1st, 1814), "before untried combinations can be brought to act in unison and harmony, that nothing should be left to chance, or as a temptation to plead ignorance of the intentions of the projector."

His solicitations were at length acceded to, and the management of the saw-mill was placed in Mr. Bacon's hands " until it shall be brought to that state that it may without risk be left to the management of others." That this was a wise conclusion we may well believe, when we find difficulties occurring in other quarters where his machinery was employed, but where competent supervision was neglected,

In reply to the Messrs. Borthwick, of Leith, February 11th, 1815, relative to the use of straps for transferring motion, he says, " at Woolwich and at Chatham, I use nothing but straps, and I could not substitute any other means. The Chatham mill has many times worked with 20 and 24 saws upon one frame, cutting 30 to 40 feet in length so as to have produced as much as 900 feet out of one log; with all that, no extraordinary tightness of strap is required." And again he says, " at the Woolwich mill, where they cut frequently with from 12 to 15 saws upon one frame through very large logs, and occasionally with 6 or 8 saws through the largest ash, they never have broken a crank nor a rod." " But," he concludes, " if the steam engine were to be let loose, I would not stand by any of the machinery."

For upwards of thirty years, Bacon retained his situation, justifying in the fullest measure Brunel's recommendation, and exemplifying the economy of securing competent supervision over mechanical combinations, such as those brought into operation at Chatham.

In 1843 I find the steam engine is reported " to be working with the original brasses ; that neither the main shaft, running fifty-four revolutions per minute, nor the other shafts, were ever broken, nor the brasses worn out," but are all in as good a state as when first put to work in January, 1814.

This is, I believe, unique: sawing hard wood being considered the "most harassing work" to which machinery can be subjected, where the curvature of the timber is to be followed. The same is stated of the "waterworks," which have to raise from three to four hundred tons of water daily."

In the case of Mr. Ellicombe, ignorance of the value of responsible supervision of unusual mechanical operations can alone extenuate, though it cannot justify, the conduct of the Navy Board towards that gentleman. In 1816 Brunel was called to France for a short time. On his return he found, to his dismay, that his resident engineer had been summarily dismissed.

"Navy Office, May 9th, 1816.
" Mr. Brunel,

"I have to inform you that we have desired Commissioner Sir Robert Barlow, to signify to Mr. Ellicombe that his services are no longer required at Chatham to superintend the works connected with the saw mills,

" We are your affectionate friends,
" R. Sapping,
" H. Legge,
" Riley Middleton."

To this peremptory order Mr. Ellicombe replied, that being the agent of Mr. Brunel, he could not conscientiously abandon the trust committed to him during his absence, or without his authority. To this the Navy Board rejoined, on the 15th May, 1816:

" Sir,

" The Commissioners of Navy have commanded me to acknowledge the receipt of your letter of the 11th of this month, on the subject of your superintendence of the works connected with the saw-mills at Chatham, in which you state that you cannot give up your charge without Mr. Brunel's authority, and I am to acquaint you that your salary will not be continued longer than the 2nd of this month, being the day on which Sir Robert Barlow stated that your services were no longer wanted.

" (Signed)   G. Smith."

In his remonstrance to the Navy Board, Brunel naturally expressed his " surprise at the nature as well as the manner " of the act. " If for so short a period," he says, " as two or three weeks, Mr. Ellicombe's exertions and labours have not been so actively and usefully employed as they were before, it is because others have not been as expeditious in the execution of the works they had to perform as I had expected." " Mr. Ellicombe's services have not been continued by me solely for the purpose of superintending the saw mill, but for directing the execution of the works in general, and for giving them the effect they should arrive at before they can be left to the management of others.

" No part of the work evinces greater proof of ability and judgment, than the manner in which the timber-lifting apparatus has been put up and brought into action.

" What remains to be fixed cannot be combined with the existing works, nor connected as it should be, unless I have the entire management of the concern, as I have hitherto had, nor unless I have the choice of instruments I think necessary to my purpose. Mr. Ellicombe being, from his superior education, liberal connections, and his uncommon acquirements, fitted in every respect, I trust that, if your honourable Board has no personal objection to him, he will be allowed to continue where he is in the character of my confidential agent in superintending my Chatham engagement until I shall have completed it.

" Waiting for your honourable Board's directions and instructions, I have the honour to be, Sirs,

" Your most obedient and most humble servant,

" M. I. BRUNEL."

" Chelsea, May 15th, 1816.

To Brunel's remonstrance, the commissioners expressed their consent to Mr. Ellicombe remaining a further time in the superintendence of the works ; but in the blind eagerness for economy, they desired to be informed " how much longer it is likely that Mr. Ellicombe's attendance, at the public expense, will be absolutely necessary." But Mr. Ellicombe had already quitted Chatham in disgust. This communication was forwarded to him at Oxford, whither he had retired, with the determination to abandon the profession for which he appears to have been so admirably fitted, and to devote himself to the church, to which he had been early destined, but from which his mechanical aptitude and superior mathematical acquirements had for a time allured him. In his reply to the Navy Board, forwarded to Brunel, but for which Brunel substituted a communication of his own, we learn how this accomplished gentleman was induced to retire from a profession in which he gave the most undoubted promise of a brilliant and successful career : not, as stated by the Navy

Board to Brunel, 17th June, 1816, "from motives of self-advantage and convenience, in having withdrawn from his employment at Chatham, for the purpose of entering into Holy Orders," but, as he himself distinctly declares, "from a conviction that, if work and genius like his (Brunel's) can meet with no better reward than has fallen to his share, it is not worth any man's while of *ordinary* abilities to exert and harass himself for such a trifling reward; and therefore, having obtained Mr. Brunel's dismissal, I return to the profession of the church, for which I was originally intended."

But what Brunel's feelings were under this deprivation, his own words only too well express. In his letter to Mr. Ellicombe, on the 31st May, after alluding to his dismissal and his reinstatement " in that situation from which he had been dismissed in a most unwarrantable and ungentlemanly manner," he says, " few, my good friend, combine with a steadiness of mind those qualifications and acquirements you are blessed with. Few, indeed, unite in that moral composition, the advantages you possess. To divert all these talents from their useful course, is to deprive your country of the benefit that must have resulted from your labours, and yourself of that reward which would ultimately have been the share of your perseverance. To the five talents you have received shall you not have to say, in return, *Ecce alia quinque?* As to myself, I must submit to the loss I cannot prevent, and which I feel most particularly at this time. Alone in the middle of the action, or at any rate, in the thickest part of it, a great deal still remains to be performed, before I can say I have closed the career in Chatham dock-yard.

" The share I had assigned to you, left me at leisure to ponder upon what came next; but now no one have I at the helm—none through whom I can convey my directions and ideas — and by the co-operation of whom I can proceed with confidence. If you still continue in your determination of returning to the Church, may you, my good friend, prove as great an ornament to it as you would have been in that most arduous career in which you leave your very sincere friend, *with one of his lights out*." Smarting under the injustice of seeing his confidential agent peremptorily dismissed from an important trust, during his short absence from the country, and only a few days previous to his return, without, as he says, allowing him the opportunity of justifying the cause of having retained Ellicombe, Brunel was induced to reiterate his complaints to the Navy Board, with no other effect than to excite an unkindly feeling towards himself.

He writes to Mr. Ellicombe : " You have had the opportunity of getting an inside view of the troubles, vexations, dangers, and chances incident to the profession of an engineer ; you are therefore completely cured of that vertigo which might at times have raised in your mind speculative ideas, and left there some regrets at not having followed the bent of your natural disposition."

The wretched economy which could have induced an important government department thus remorselessly to lop off the right arm of him to whom the country already stood so largely indebted,—or, as he himself expressed it, to put "one of his lights out," must have been poor indeed ; and the conception which must have been entertained by that department, of the necessity for skilful supervision, in the erection of works of such magnitude as it had been called upon to sanction, must have been of the most circumscribed and imperfect character.

One word more relative to Mr. Ellicombe, or *Ella*combe, according to the early orthography now adopted. However much the Church may have gained by the accession of an active and earnest minister, it is certain that the civil engineering profession lost a conscientious and accomplished member.

Although removed from the busy haunts of men, and devoted to high and holy objects, Mr. Ellacombe had still recognised the benefits which a knowledge of the physical sciences is so well calculated to confer.

Of a highly respectable family in Devonshire, directly descended from that Sir Hugh Myddelton, who, like the last Duke of Bridgewater, expended all his means in the realisation of a great and beneficent idea*, Mr. Ellacombe was early destined to follow the steps of his father and grandfather, and to enter the Church. Already had he graduated at Oriel College, Oxford; but inheriting with his distinguished brother, General Ellicombe, R.E., many of the engineering qualifications of their ancestor, he was unable to resist the impulses of his nature, and every hour which he could command from his more serious studies was devoted to mechanical drawing and the construction of models.

While still at Oxford, an opportunity was afforded him of an introduction to Brunel, and to Brunel he submitted the result of his stolen hours. The delicacy, accuracy and beauty of the workmanship at once secured the favourable opinion of the great mechanist.

---

* Sir Hugh constructed, at his own cost, the celebrated aqueduct called the New River, by which, through a distance of sixty miles, he conveyed the first wholesome supply of water to London, thereby conferring upon the metropolis of his country an incalculable blessing. That noble work was commenced in 1608, and completed in five years.

This was enhanced and confirmed by the simplicity of
young Ellacombe's manners, and the superiority of his
general attainments; and he soon found himself, he
scarce knew how, installed in Brunel's office; from
whence the transition was natural to the important
position of his confidential assistant.

Although, in consequence of the conduct of the
Navy Board to Brunel, he was led, as we have seen, to
abandon the profession for which his natural endowments so well fitted him, it is some consolation to know
that his mechanical and constructive talents have not
been lost to the country.

As curate and vicar of Bitton, in Gloucestershire, for
upwards of thirty years, and subsequently rector of
Clyst St. George, in Devonshire, where he has within a
few years entirely rebuilt the parish church, and erected
spacious school-houses, he offers another illustration of
the well known adage of the poet,

" Naturam expellas furcâ, tamen usque recurret."

I cannot conclude this short notice of a most amiable
and gifted gentleman, without recording the opinion of
a stranger who, under the title of " The Bristol Church-
goer," thus describes the value of some of Mr. Ellacombe's labours.   Writing in 1849, he says —

"The Vicar of Bitton is one of those men who, if
you placed him in the desert of Arabia, would, I believe,
have half a dozen churches up about him in little
more than that number of years.   I'm afraid to say
how many he has built in the parish of Bitton, which
was once as bad as Arabia ; but I think I am correct
in stating that he found it with one, and that he has
managed to add four or five others, and by the time

he is gathered to his fathers, as many more will, I expect, stand as monuments of his untiring, his unconquerable zeal. Where he gets or got the money for them all, Heaven knows, I don't; but I should say he must have been a most intrepid beggar, and indefatigable man to do what he has done. He restored the mother church; he rebuilt Oldland; perched a pretty new chapel on Jefferie's Hill; and planted another amongst the coal pits of Kingswood.

"He rubs on, enlarging the borders of the Church, while others are squabbling about her. Erecting altars, while others are fighting about turning their faces toward, or from them."

We should form but an incomplete estimate of the real value of those beautiful arrangements at Chatham, if we permitted our views to be limited to the process of sawing only, economical as that undoubtedly was; the fact being, that the Chatham works really formed a *new era in the economical management of timber.*

However painful it is, on the one hand, to record the nature and amount of vexation to which the sensitive mind of Brunel was subjected, it is, on the other, gratifying to find that there were those of superior attainments and position who had formed a true estimate of his genius and services.

Sir Joseph Yorke, in a letter to Brunel, dated 16th January, 1821, referring to the interest which he took in Brunel's works, and the circumstances which enabled him to forward the successful operations at Chatham, while he had a seat at the Admiralty board, adds:—

"Your talents and ingenuity, united to profound mechanical knowledge, have placed you in the envied

situation of the projector of some of the most complete
apparatus that has ever appeared, even in this mechanical country."

" That you may live to reap the harvest of such well-
earned success, as well in pocket as in fame, is the
sincere prayer of

" Your sincere admirer and friend

" J. YORKE,

" Vice Admiral."

Kind Sir Joseph Yorke; how little he knew of those
dark clouds already gathering on the horizon,—precur-
sors of the storm about to overwhelm the fortunes of
his friend!

There were others, also, qualified to estimate the
value of Brunel's labours, who did not hesitate to
express their opinions as to the sort of justice he had
received at the hands of government.

We have already seen in what light Lord Spencer
viewed the concession made by the Admiralty, for the
invention of the block machinery; nor was Lady
Spencer less indignant. This lady, no less distinguished
for her high social position than for her intellectual
attainments, her moral virtues, and the sympathy for
truth and goodness and genius which endeared her to
all who had the privilege of her friendship, took from
the first hour of her acquaintance with Brunel, and
when his talents were scarcely recognised, the deepest
interest in his career. She now, in his time of trouble,
addressed to him one of her earnest letters.

" I am in a sad hurry, or I should wish to tell you
of a great desire I have, that you should just now make
a very short and very clear statement of the case to

the country; both on the block-machines, and on the saw-mills, and their appendages at Chatham. I want you to draw up such a statement, from the documents which you must possess, as might be written on a *large card*, so that your advocates might instantly place before *an ignorant age the wonderful result*, and impress the most stubborn, with a plain matter-of-fact which must place your pretensions on the most self-evident grounds, and so as to silence all other claims. I wish I could better explain what I mean; but I think you may conceive my intentions.

" Ever yours in haste

" LAVINIA SPENCER.

" Wimbledon, Saturday."

In 1854 these valuable works were burnt down, but they were too important to admit of any delay on the part of government in their re-establishment. They were therefore rebuilt in 1855, and the only changes, as reported by Captain Superintendent Goldsmith to Lord Clarence Paget, March, 1861, were "alteration in the mode of driving, and the speed of the frames, which has been increased from 80 to 100 revolutions a minute." To this has been added, "two steam frames and three circular saw benches." *

We thus see how far the value of Brunel's genius has received the testimony and seal of time. It is very certain that no mind shed more light on the practice, if not the principles, of constructive engineering. No hand performed more labour, no life rendered more

---

* I am again indebted, through Sir Benjamin Hawes, K.C.B., to Lord C. Paget for a report upon the present state of the mills, which has enabled me to place their continued connection with Brunel upon record.

consistent and essential service to the mechanical requirements of the country than his did ; and it is therefore with feelings of deep regret and sympathy, that we are called upon to contemplate the anxieties and disappointments to which he was now subjected.

CHAPTER IX.

1809—1819.

ORIGIN OF THE SHOE MACHINERY — NEGLECTED CONDITION OF
THE BRITISH ARMY AND NAVY — DESCRIPTION OF SHOE MA-
CHINERY — ENCOURAGED TO ESTABLISH A MANUFACTORY OF
SHOES — TERMINATION OF THE WAR, AND LOSS INCURRED, 1814
—LETTER TO MR. VANSITTART, 1819.

WHILE the works at Chatham and at Woolwich
were in process of erection, the saw-mills established at Battersea by Brunel, in connection with Mr.
Farthing, in 1808, had been brought to considerable
perfection. Applications were also being received from
private individuals to supply saw-mills. From his
Grace the Duke of Athol, I find a kind invitation to
Dunkeld, where his Grace had proposed to erect mills.
At Battersea, circular saws, from seven to eighteen feet
in diameter, were performing work never before contemplated, with a rapidity and precision altogether un-
equalled, and entirely independent of the mechanical
supervision of the projector; while the establishment
under ordinary, but faithful management, promised to
secure an ample provision for his declining years. Why
it did not so prove, it will be our painful duty presently
to narrate.

The idea of applying machinery to everything which
was calculated to contribute to the utility, the economy,
and the comfort of life, stimulated, as we have seen,

Brunel's mind to devise the means of sewing and hemming. A higher motive now urged him to seek to supply,
by machinery, boots and shoes for our soldiers, which
should be independent of the shoemaker's wax and
thread, and the contractor's cupidity and knavery.

He had witnessed at Portsmouth, in 1809, the disem-
barkation of some remnants of that gallant band which,
under Sir John Moore, had so gloriously sustained the
honour of the British name, through difficulties and
privations at that time almost without parallel. He
had learnt how greatly the want of shoes had contributed to the losses which the army had sustained, and
his kindliest, deepest sympathies, were at once enlisted
as he looked upon those victims of cupidity and neglect,
dragging their mutilated limbs along, shoe-less ; or with
lacerated feet enveloped in filthy rags bound round with
knotted strings.

As the necessities of our navy had stimulated his
constructive genius in 1797, so the sufferings of our
army now awakened his philanthropic spirit to devise
a remedy for a great and crying evil.

For six years the average annual expense of shoes
supplied to the army amounted to 150,000*l.* ; and yet
it is on record that those procured at such a cost had
been found inadequate to *one day's march :* indeed the
disgraceful manner in which the army contracts had
been executed during the Peninsular war, had become
the great scandal of the country. Clay was introduced between the soles to give weight, and therefore
the appearance of great strength of leather. This,
while in dry weather it produced extreme heat to the
foot, in wet weather dissolved and left the unfortunate
victim of cupidity in the most distressful condition.
Little less disgraceful to the country, or perilous to the

K

interests of the service, was the character of the sail cloth permitted to be supplied to the navy.

The cloth, like the leather, was also paid for by weight, and was composed of the most worthless materials, to which the necessary qualification was given, says Lord Dundonald, in his deeply interesting and instructive *Autobiography of a Seaman*, by "a composition of flour and whitening, so that the first shower of rain on a new sail completely white-washed the decks."

"Of so flimsy a nature were the sails," adds his Lordship, "that when the composition was washed out, I have observed the meridian altitude of the sun through the fore-topsail, and by bringing it to the horizon, through the foresail, have ascertained the altitude as correctly as I could have done otherwise:" and he further says, "the enemy distinguishes our ships of war from foreign ships by the colour of the wretched canvas, and run away the moment they perceive our black sails rising above the horizon ; a circumstance to which they owe their safety, even more than to the open texture of the sails." *

If the great bulwarks of the country were thus neglected, we cannot wonder that the internal economy of the service was utterly disorganised. Lord Dunfermline in his "simple and scrupulous exposition" of the services of Sir Ralph Abercromby says : — " The troop ships, with very rare exceptions, were in a most wretched condition, deficient in anchors and stores of every kind ; the decks so leaky, that when it rained the men were constantly wet ; they were so much crowded, that the soldiers, with no other bedding than blankets, were obliged to lie on the deck." †

* Autobiography of a Seaman.
† Memoir of Sir Ralph Abercromby, K.B., by his son James Lord Dunfermline.

In 1810 Brunel devised his shoe machinery, and in February 1811, he had his patent enrolled. The soles were fastened to the upper leather or closing by means of "metallic pins or nails;" leather, bone, whalebone, or catgut might also be used. The upper leather was well stretched upon a cast iron last to which it was held close, by metal clamps ; the clamps being kept in place by pawls acting upon joints. The soles were cut to the proper size on an iron frame ; and the inner sole being laid upon the last, and the projecting edges flattened down, it was ready to receive the sole. The whole was united by nails, by the operation of a vertical rod attached at the top to a lever worked by a treddle, and at the bottom to an awl and a hammer. By means of a guide, a moveable frame carried the last round with a uniform motion, presenting the rim of the shoe to the base of the rod. The fall of the rod forced the awl through the leathers, and at the same time struck in the nail which had been dropped into the hole previously made by the awl as the rod rose. Nothing but the usual trimming and polishing was required to complete the shoe before it was taken off the last. A special machine was added for making nails ; as also a process for " rendering leather durable," more particularly applicable to that " used for making the soles of shoes and boots." For this a patent was secured in September 1814. The process consisted "in studding the leather with small nails or pins, after it had been previously saturated with a composition of neat's-foot oil mixed with tar in the proportion of nine or ten to one." In the same patent reference is made to instruments devised for making the nails, and driving them into the leather, which formed part of the shoe machinery.

The manufactory was divided, according to an entry

ın Brunel's journal for 1814, into sixteen processes, and the sizes of shoes into nine.

The operations consisted of punching, tacking, welting, cutting or trimming, nail-driving; and the number of men employed was twenty-four.

Fixing the heel	.	.	.	.	.	.	4
Rubbers	.	.	.	.	.	.	10
Parers and trimmers	.	.	.	.	.		10

$$24$$

The number of nails I find given for two of the sizes, viz. No. 9 and No. 7. No. 9 required for

Soles.	Studding	. .	177		
	Tacking	. .	91	383	
	Long nails	. .	115		555 : and for
Heels.	Studding	. .	140		a pair 1110.
	Long nails	. .	32	172	

The prices of these shoes and boots were as follows:—

						s.	d.	
Common shoes.	.	.	.	.	.	9	6	a pair.
Water boots	.	.	.	.	.	10	6	,,
Half boots	.	.	.	.	.	12	0	,,
Superior shoes.	.	.	.	.	.	16	0	,,
Wellington boots	.	.	.	.	.	20	0	,,

and this at a time when iron was 34s. a cwt; leather 1s. 11d. per lb.; and wheat from 126s. 9d. a quarter in 1812, to 74s. 7d. in 1814.

The machinery possessed all that symmetry for which Brunel's works are remarkable. The simplicity of the construction also enabled it to be worked by invalid soldiers, many of whom were cripples; yet the superiority of the shoes, as regarded durability, finish, and cheapness, was unexampled. In consequence of the favourable opinion expressed by various individual members of the government, Brunel was induced in 1812 to prepare machinery sufficient to supply the whole of our forces with shoes, and to secure premises suitable to so important a branch of manufacture.

The shoes proved all that could be desired : 400 pair were produced daily. A large order was issued by the government, which was completed within the time stipulated ; but, unfortunately for Brunel, when everything was in full activity, and the workmen had become familiar with their work, the war had come to an unlooked-for termination ; the government no longer required the aid of the shoe machinery ; while Brunel, relying too implicitly on the moral obligation by which he believed the government to be bound, continued to incur the heavy liabilities connected with a manufactory in full operation. The consequences were serious. A large stock of shoes, for which there could be no demand, was accumulated, and financial difficulties arose from which Brunel was unable to extricate himself.

In 1819, the 22nd February, he addressed a letter to Mr. Vansittart, the then Chancellor of the Exchequer, in which he describes, first, the condition of the unfortunate soldiers, as he had seen them at Portsmouth, and then adds :

" I was prevailed upon and induced, therefore, to turn my attention towards supplying the deficiency ; the great trouble and expense of which I took entirely upon myself. I invented and prosecuted a plan for making, by machinery, military shoes of a greater durability than those previously used in the army.

" As soon as the machinery I had invented was in a state to work, I applied to the Invalid Department for disabled men whom I proposed to be exclusively employed in it. When my plans were known to the principal officers in that department, directions were given to afford me every assistance I might require to accomplish my object.

" Having once the opportunity, when H. S. Highness the Prince of Orange visited the Royal Arsenal at Woolwich, in March, 1813, to mention to several of H. M. Ministers then present, that I was engaged in forming an establishment for making military shoes, I was commended for so laudable an undertaking.

" It was admitted that the shoes sent abroad were proverbially bad, and several officers then present added to these observations facts that excited a great deal of interest in the success of the scheme. My Lord Castlereagh, addressing himself to me, said, ' When you are ready to make any number of them, let me know.' I have since complied with his Lordship's directions, but have received no answer.

" The Commissary-in-Chief, sensible of the importance of the undertaking, gave it every encouragement in his power, in taking all its produce in its infant state. Shoes were supplied to the 13th Regiment, and the report of them was favourable; but objections were made on the supposed impossibility of mending them.

" In actual service in the field, it is found that the mending cannot be practised even with the common shoes; consequently the *intrinsic value of these shoes consists in their first durability*, and in the quality they possess beyond the ordinary shoes.

"In the course of 1814, H. R. H. the Commander-in-Chief, being informed of the progress I had made, did me the honour to visit my establishment. H. R. Highness and the officers of his suite, passed the most flattering encomiums on the utility of the work. It was stated on that occasion, that shoes sent for the service of the army had been found *not to last one day's march.*

"I applied, from time to time, to the several departments of government for a more decisive protection of my establishment. A decisive answer was always delayed, and I could not but proceed.

"The Commissary-in-Chief continued to take some shoes from me till the period came when, in March, 1815, a new continental war broke out; but no shoes were as yet ordered by the Treasury, and the reason assigned was, that there were upwards of 30,000 pairs in store.

"These repeated instances of acknowledgment of the importance of my undertaking led me on, and operated as a conviction that I could not have rendered a greater service to the country than I had already done, since it is well known that shoes are the first article in military service. My exertions were unremitting till the beginning of July, 1815, when, from apprehension that I might find myself too deeply involved, I addressed, on the 6th of July, H. R. H. the Commander-in-Chief, stating that I was incurring an expense of 200*l.* a week; that I could proceed no farther, as there was no assurance of protection from Government. H. R. H. came immediately to see my works for the second time, being accompanied by Lord Rosslyn and Lord Palmerston.

" H. R. H. inquired minutely into everything; saw that
every attention had been paid to the comfort and health
of the men, and was shown the stock of shoes I had on
hand; when H. R. H. was pleased to observe to their
Lordships, ' We must take this from Mr. Brunel.'

" A few days after, H. R. H. recommended the soldiers
of the army in France should be supplied with one
pair of shoes gratuitously, and directing that the shoes
from my establishment should be taken in preference.
My Lords of H. M. Treasury directed the shoes thus
ordered, to be taken from those in possession of the
Commissary-in-Chief; and as none were purchased from
me, the opportunity of relieving me proved of no
avail.

" At the end of July, 1815, I waited, in consequence
of this disappointment, on Lord Palmerston, in order
to know how I was to be relieved from the heavy inconvenience I was labouring under; being, as I stated it,
in advance 5,000l. His Lordship gave me the assurance
that I should not be deserted, and stated that I should
receive a more decisive answer. I renewed my application a week after; when I was directed to apply by
letter. I accordingly addressed his Lordship, observing
that I was incurring a weekly expenditure of 200l.,
which advance I could support no longer. I farther
desired to be informed whether it was safe in proceeding farther, begging for an immediate answer.

" The delay in answering me, under such circumstances, led me to anticipate a favourable result. In
this dilemma, I found myself compelled to make an
additional sacrifice, as I might show by my banker's
account, of the sum of 4,500l., which the Right Hon.
the Master General of the Ordnance had granted me
for my services in the Royal Arsenal at Woolwich.

" Receiving no communication from Lord Palmerston, I repeated my application ; and on the 24th of October (1815), I was honoured with the following answer, intimating 'that his Lordship is so persuaded of the superiority of the shoes made at your manufactory over the common shoes the army is supplied with, that had the demands of Government called for new supplies, you should have been, at his recommendation, applied to for such quantity as you might furnish.'

" Thus, you will observe, a decision was deferred until a favourable decision was of no use ; for peace had now taken place. It is true that Government took at last some little notice of my claim, and, on June 25th ensuing (1816), ordered one half of the stock of shoes then in my hands should be purchased at a diminished price. I was obliged by necessity to sell them at the price then reduced, and ultimately to get rid of the remainder at less than two-fifths of the prime cost. But this very purchase, while it shews an acknowledgment of my claim, was, as you see, an addition to my loss.

" But, sir, as I said before, the peace was at that time come, and the utility of my establishment lost sight of, and I might say, *the recollection of it gone.*

" The loss alone has fallen upon me, a loss amounting to upwards of 3000*l.* sterling ; besides the incumbrance of premises, for which I must pay 400*l.* per annum, for they are still upon my hands.

" For this loss, sir, incurred in this way and on the ground which I have detailed to you, I solicit a compensation which I feel I have a title to. The loss was incurred on an object for Government ; the loss was incurred, as it were, with the very knowledge of Government ; for I never should have gone on absorbing my own pecuniary resources, the fruits of my past services, unless buoyed up with the assurance and by the hope that Government would support me. I should never have gone on if Government had given me a negative to my repeated solicitations on this subject.

"I do therefore trust that I shall not be held to ask too much, in asking that the Lords of H. M. Treasury will be graciously pleased to take this statement into their favourable consideration, and order to me a compensation of my losses, chiefly of the sum of 4500*l.*, being that part of my former earnings from Government, which, as I before stated, I have absolutely sunk, together with much more of my own property in the establishment, of which I have now troubled you with the detail.

"I have the honour to be, Sir,

"Your obedient servant,

"M. I. BRUNEL.

"To the Right Honourable Mr. Vansittart.
  February 22nd, 1819."

Finding that it was hopeless to expect any consideration from those in authority at home, Brunel entered, January, 1819, into negotiations with the Prussian Government for the supply of shoe machinery. On the 25th of May he forwarded a detailed specification, not only of the machines to be supplied, but of the extent of buildings which should be assigned for the effectual working of the different departments of a great national establishment : there is no evidence however, that the Prussian Government adopted Brunel's plans.

For many years the practice in the Prussian Service

has been, for each regiment to employ its own soldiers who understand the trade, to make boots and shoes for the corps.*

* A manufactory for shoes has, I understand, been recently established in Haverhill, U.S., by machinery. All the stitching is done by sewing machines.

The account states, " that one of the most curious parts of the machinery is that allotted to the formation and driving of the pegs by which the soles are united. The peg is cut from a strip of wood; the awl forms the hole to receive the peg; and the peg is driven; all by one operation, and so rapidly, that two rows round the sole of a shoe are driven in twenty minutes. By means of this machinery, twenty-five men produce six hundred pairs of shoes per day."

CHAPTER X.

1814—1819

FELLOW OF THE ROYAL SOCIETY, 1814 — DOUBLE ACTING MARINE
ENGINE, 1814 — RECEPTION AT MARGATE — MR. HYDE — PLAN
FOR TOWING VESSELS, 1816 — KNITTING MACHINE, 1816 — TIN
FOIL, 1818 — CONNECTION WITH MR. SHAW — STEREOTYPE
PRINTING, 1819 — RETROSPECT OF THE ART OF PRINTING —
BRUNEL'S IMPROVEMENTS IN THE PRINTING PRESS, 1819.

ON the 23rd March, 1814, Brunel was elected a
Fellow of the Royal Society, as a tribute which
the highest scientific body in the kingdom was willing
to pay to original and creative genius; and in recognition of the practical influence which mechanical invention had exercised upon science itself. It is very
certain that the value attached to machinery has progressed with the advance of civilisation, and that those
countries which have shown themselves most willing
to encourage its application, have secured to themselves
advantages which others will in vain hope to realise.
In 1832, and under the presidency of H.R.H. the Duke
of Sussex, Brunel was elected a Vice-President, and as
far as I am able to discover, he is the only foreigner
who has ever filled that honourable position.

In this year (1814), Brunel made his first experiment on the Thames, with a double acting marine steam
engine. Having accomplished his voyage to Margate,
he was desirous of obtaining accommodation for the

night, but this was not easy. So strong was the prejudice which this new mode of communication excited
in the minds of the inhabitants, particularly those connected with the sailing packets, that, blind to their
future interest, they threatened personal injury to
Brunel, and the landlord of the hotel absolutely refused
to provide him with a bed.

In a letter which I received from Brunel dated the
10th September, 1836, when he was engaged in advis-
ing the survey for the proposed railway to Ramsgate,
he says :—" To-day, by mere chance, I am at the *York
Hotel*, for my arrangements were at another hotel. It
is at *this same hotel* that in 1814 I was refused a
bed because I came by a steamer, and every one of the
comers met with a very unfriendly reception. If they
knew at this moment that I come to carry off the
cargoes of the steamers to Ramsgate I might probably
share the same fate." *

* A case somewhat similar is recounted by Dr. John Foster, in
his account of the "*grand remonstrances*," where Mr. Hyde, afterwards Lord Clarendon, found himself at York, shortly after his
patriotic exertions "to stem a tyranny that passed over the land
like a net-work, and which, excepting only its agents and projectors,
did not permit a single class of the community to escape." "There,"
says Foster, "he became curiously aware of the impression which
his exposure of the *Council of the North* had made in that ancient
city." A lodging had been provided for him by one of the King's
servants, with which he was well pleased — and his servant was
made welcome by the mistress in a room below. "They were
sitting together there quite pleasantly," says Hyde. "She (the
mistress) asked him (the servant) what his master's name was, which
he told her. '*What !* ' said she : ' *That Hyde that is of the House
of Commons ?* ' and he answering yes, she gave a great shriek, and
cried out that he should not lodge in her house ; cursing him with
many bitter execrations. Upon the noise, her husband came in ;
and when she told him who it was that was to lodge in the chamber
above. he swore a great oath that he should not ; and that he would

Yet a year had scarcely elapsed after the introduction of steam communication between London and Margate than we find 32,500 persons taking advantage of this dreaded innovation, a number which in 1830 had increased to 95,000, and ultimately to 2,000 a day; the steam boat service being regularly performed throughout the year, in place of being limited to four months.

On the 11th of October, 1816, Brunel proposed to the Navy Board a plan for towing vessels of war by means of steam, and negotiations were entered into with the proprietors of a steam packet called the "Regent," which had just completed *her season* of going to Margate, to enable Brunel to make the necessary experiments. The demand for the *hire* of the vessel only was 52*l*. 10*s*. a week, "because," say the proprietors, "the service being extraordinary in which she is to be employed, no price for hire can indemnify the owners against the casualties of extraordinary wear and tear, and no insurance can be made to cover the risk." Brunel's proposals were at first accepted, and certain alterations in the vessel sanctioned; but the Navy Board, after nearly six months' deliberation, abandoned their intention, and disputed the charge for the altera-

rather set his house on fire than entertain him in it. He knew him well enough; he had undone him, his wife and his children!' Such was the servant's account, with more oaths and slamming of doors than may here be dwelt on; and for which, on Mr. Hyde's resolving nevertheless to wait till morning to try and find out some rational explanation, the next day brought reason enough. The man of the house had been an attorney in the Court of the President and Council of the North, in great reputation and practice there, and thereby got a very good livelihood, with which he had lived in splendour; and Mr. Hyde had sat in the chair of that Committee, and carried up the votes of the Commons against that Court to the House of Peers, upon which it was dissolved."

tions already made in the vessel. Brunel was hurt and disappointed. He had hoped that a favourable opportunity would be afforded him of realising ideas which he had long entertained, and which " a series of very extensive experiments," he said, " had convinced him of the practicability of accomplishing." Unfortunately I can find no record of the experiments — the only explanation of his views appears in a letter to the Admiralty, 2nd May, 1817, where he proposed to attach to a vessel purposely constructed " such machinery as would render it capable of heading very heavy seas and a gale of wind, and fit to carry cables and anchors to the assistance of ships in distress and for any service equally important." And he adds, " since July or August last I have been at considerable expense in obtaining self-navigating models, which have been made at my own cost, at my own works here, and under my own immediate direction, for the particular purpose of exhibiting in a satisfactory and comprehensive manner what I am now advancing." The Admiralty were however inexorable, and the advantages which Brunel offered were, for the time, lost to the country.

In September, 1816, Brunel took out a patent for a *tricoteur*, or knitting machine, by which the whole of a stocking could be executed in one piece. Without drawings it would be impossible to convey any correct idea of the ingenuity of the machine, the largest wheel in which did not exceed sixteen inches in diameter. It might be worked by hand, and would occupy but a small space in a room.

This machine appears to have become the property of M. Ternaux, of Paris, and never to have come into operation in this country.

But it was not with the useful only that Brunel's

mind was occupied. The adornments of life had for
him a great charm. We have seen how strongly the
artistic feeling prevailed, and what pleasure he derived
from the contemplation of the beautiful, whether in
form or colour. It was therefore only natural that he
should be attracted by the novel effect produced by
the crystallisation of tin, when cooling from a state of
fusion. Having subjected the process to a variety of
well-directed experiments, he, in 1818, and in conjunction with Mr. Samuel Shaw, a private friend, took
out a patent for a peculiar preparation of tinfoil, which
should combine with an elegant form of crystallisation
the ever-varying charm of colour. The process con-
sisted in first melting sheets of foil about $\frac{1}{600}$ part of an
inch in thickness, by spreading them upon a plate of
iron heated to a temperature a few degrees under that
at which tin melts. When the surface was rendered
perfectly smooth, and all air carefully excluded from
beneath the sheet of tin, the necessary additional
degree of heat to produce fusion was applied by means
of a flame of gas, conveyed through flexible pipes, " by
which it was rendered as manageable as a pencil."
The result was a " large, varied, and beautiful crystallisation." The action of acids was found necessary to
complete the process, and the surface was then protected by a varnish more or less transparent. The
popularity with which the article was received in its
application to boxes, columns of lamps, urns, &c., led
speedily to numerous encroachments upon the patent,
which inflicted a large amount of injury on the patentees
in protecting their rights. So vexatious, indeed, were
these attacks, that Mr. Shaw, who seems to have been
a gentleman of education, courtesy, and probity, but
whose lively imagination readily suggested either a

success which was scarcely warranted, or a loss which he was unwilling to contemplate, was frequently tempted to abandon altogether the speculation, while Brunel was only stimulated to further efforts. So early as May, 1818, he writes to Brunel: "I am so sick and tired of the vexatious occurrences and disappointments at Battersea, that I should be glad to get rid of the whole concern altogether. Others are making fine profits by the invention applied to tubes alone, of which as I told you, 500 were supplied by one man. As to going out to the works, I must give it up. The walk is too much for me, and, upon my honour, I cannot afford 12s. for the ride. I feel no inclination to incur fresh expenses by going to law; indeed the uncertainties of the whole concern do not appear to me to justify it."

In June things looked better; the manufacture had been introduced to "inkstand makers, cabinet makers, coach makers, urn makers, lamp makers, stove makers, consumers of tubes, &c., and exported to Calcutta and Madras;" and Mr. Shaw closes a letter to Brunel thus:

"When 20 per cent. on the sales of our manufactured plates shall amount to 2400l., I'll sing out 'Brunel for ever!' and be your slave as much as I am now,

"My dear Sir,
"Your sincere friend,
"S. S."

In December, 1818, Brunel introduced another application of tin foil in the form of metallic paper, which he again secured by patent on his own account, both for this country and for France. M. Chaptal and the Society of Arts of Paris, as well as the "Journal du Commerce," spoke in the most favourable terms of the

discovery. Parisian taste had already given *éclat* to
the first patent, and it was fully anticipated that this
second introduction would speedily find its way from
Paris throughout the capitals of Europe. The article
was also introduced to ornament some of the rooms of
the Pavilion at Brighton. Still, notwithstanding the encouragement which this novel and beautiful invention
received, the want of business arrangements again interfered to mar its development, and to rob the inventor
of his anticipated reward. Mr. Shaw complains, " that
with a tenant already in the possession of nearly half
the premises at Battersea, and another occupying a
considerable portion of what remains, sufficient accommodation was not provided to enable the manufactory
to meet the present demand, to say nothing of what
might be expected from a fast and very extensively
increasing necessity in this respect." It must also be
added, that the treatment which Brunel appeared to
meet with from the speculators and tradesmen of the
country with whom he became connected, made him
unwilling to hazard an increased outlay in the shape
of rent. Indeed, Mr. Shaw also seems to have entertained a becoming horror of those vampires of commerce.
" I dread," says he, " both the encouragement and the
villany of my country." As a man of business it is
curious to find that Mr. Shaw had neglected to adopt
the most ordinary and best recognised mode of pro-
secuting the business part of the establishment with
which he was so intimately connected, and which at
times created in his mind so much anxiety. In the
beginning of 1819 he stated, in a reply to Brunel, that
he had nothing to recommend for the extension of
their business, but an active traveller to exhibit specimens of the works. To this Brunel rejoins, " You

will find all through my correspondence that I have considered that plan, and have pressed its execution for a long time as the only way that can insure the success of this as well as of any business. Indeed, it is more urgent at the first onset than when a concern is established. Having had no one capable of filling it out of doors, nothing is doing in a way likely to meet the wishes of the trade; no plan laid down; and the most valuable time is wasting away. Many of those who have manifested a wish to have the article, have heard nothing of us since; and, from the want of that intercourse that is the base of business, most particularly of a new one, nothing is going on. If you would have come only once a week, you would have seen how matters went on. It is impossible to trust everything to correspondence. I have kept only the colours in order to prepare the article as it is required. Others can sell plates, whereas the Battersea concern is left to itself! How can you expect any satisfactory result from such mode of proceeding? And yet, as Ackermann says, much may be done; but not in the way we do it.

"In your note of the 8th, you state that as far as your assistance in relieving me from the trouble of the cash department can be availed of, it will be at my command. Now, my dear sir, it is six months, if not more, since you have put your foot on the premises of a concern in which you are so deeply interested. Don't think, in making this observation, that I do it with any degree of unfriendly feeling, but think only how it is possible to convey to you by letter anything connected with it. You have had accounts; the books are in good order; the manufacturing part is, I think, complete; but the main point is left to chance." Brunel then goes on to say that there should be some one appointed that would be "the link between" Mr. Shaw and himself — one who might see Mr. S. every day in the city — the progress of the demand being always sufficient information for him ; but he concludes : " In the present state I would rather do as you propose, wind up the business in a way that will protect it against those claims that are upon it. I would rather sacrifice the whole at once than suffer things to go on in the present way." And thus it was, that another valuable discovery, and one which promised to become highly remunerative. was suffered to slip away from his hands, from the want of the most ordinary means, which almost every manufacturing tradesman in the kingdom was in the habit of employing.

In 1819 Brunel made a variety of experiments on the best materials for forming stereotype plates, and the " metallic compound of which the stereotype " should be made. " Nights as well as days were devoted," says Lady Hawes, " to the solution of the chemical and mechanical difficulties which presented themselves." Before the year had expired, the problems were solved, and a patent was secured " for accelerating the printing of daily papers," and for the " making of stereotype plates in general in an easy and expeditious manner."

Moveable types could only be kept standing by privileged printers for such works of constant demand as bibles and prayer books ; but this was done at great cost and inconvenience. Efforts were therefore made, in the commencement of the eighteenth century, both in France and Holland, to supply some form of stereotype printing. In Edinburgh a goldsmith, of the name of Ged, invented (1725) a method which was admitted to be highly economical ; but, from the ignorance which prevailed amongst workmen and others interested in the old system, it failed to afford him any return : nor was it until fifty years later that Mr. Tilloch endeavoured to revive the invention, which from various circumstances was again laid aside. The ingenuity, perseverance, and influence of Lord Stanhope, a few years later, ultimately succeeded in bringing the art into very general use. Still there was ample room for improvement, not only in the preparation of the moulds, and in the composition of the types, but in the mechanical arrangements of the process of printing.*

The chief objects sought by Brunel were " to enable editors of daily papers to print their papers in a very expeditious way by the aid of duplicate plates or forms, in lieu of setting up two, three, or four times, whereby advertisements and other information might be sent to press at the latest hour."

I may, perhaps, be pardoned for venturing to break the thread of my narrative whilst I indulge in a slight retrospective glance at the earliest record of the art of printing ; an art which, above all other arts, has most advanced the interest of civilisation.

Let us then picture to our minds Guttenburg travelling from Strasburg to Mayence with his original, but still immatured invention. We watch with interest his introduction to that sagacious citizen Fust, or Faust, who possessed not only the enlarged intelligence to comprehend, but the worldly wisdom and the pecuniary means to promote the great work thus begun. We see

---

* The paste which Brunel employed for receiving the impression of the moveable types was composed of " pipe-clay seven parts, chalk or burnt clay pounded very fine twelve parts, and starch one part *in bulk.*" For the metallic compound he employed the alloy of bismuth 10 lbs., lead 6 lbs., and tin 4 lbs., heated to 400°.

the hope of gain strengthening in the mind of the merchant as the mechanical skill of his servant, Peter Schöffer, ultimately enables Guttenburg to perfect his invention. We understand the magnitude of Faust's anticipations, when, in the joy of his heart, he adopts Schöffer as his partner, and gives him his beloved daughter in marriage. And we follow him to Paris, where, in secret, he applies his art to the imitation of the MS. copies of the Bible, whereby for sixty crowns he is enabled to sell, for a large profit, that for which the scribes received 500. We witness the amazement of the people. We behold their wonder increase, as they discover that each copy is a fac simile of the other. We watch their puerile and hopeless efforts to discover the cause of this departure from recognised usage, and we listen to their superstitious fears and dark insinuations. We hear the cry of magic arise. Through the instrumentality of the devil only could such feats be performed. We see the informations laid before the magistrates, and Faust denounced as a magician. The red ink employed in his illustrations is his blood— there can be no doubt of it—it is so beautiful, so brilliant. We hear him solemnly pronounced "in league with the infernals." We see the bonfire prepared, and we contemplate the struggle with which acquisitiveness yields its prey—the reluctance with which the grasping hand is opened, and the bitterness with which the merchant's cherished secret is at length revealed ; and lastly, we behold the parliament of Paris, in all the plenitude of its power, magnanimously discharging him from all prosecution in consideration, not of the importance, but of the marvel of his invention.

But to return. The advantages which Brunel's improvements offered, speedily excited the attention of the leading journals, and negotiations were entered into with the " Times " and the " Courier " for the use of the patent.   When we find that upwards of *three hundred thousand* moveable pieces of metal were required in the ordinary process of printing the " Times," we can understand with what interest any suggestions emanating from so accomplished a mechanist as Brunel were received, by which the necessity of upholding such an enormous stock of type would be removed, while the facility of multiplying impressions would be greatly increased.   This connection with the printing house naturally suggested to Brunel's mind the means of improving the printing machine itself.   With all that enterprise, liberality, and intelligence for which the leading journal has ever been distinguished, negotiations were immediately opened with Brunel, and on the 11th of December of this year (1819), an agreement was concluded with him by the proprietors of the " Times " for the use of certain improvements ; but in consequence of Brunel having subsequently suggested important additions, including a " revolving or cylindrical press," a long and unsatisfactory discussion ensued, and in 1821 the original agreement was cancelled.   Whether it was that Brunel failed to realise the expectations he had raised, or that in the improvements introduced by Cowper, and subsequently by Applegarth, on the machine of König, Brunel's was really superseded, I have not learnt.   In concluding these observations relative to printing, it may not be uninteresting to recall to mind the chronological order of those inventions to which civilisation stands so largely indebted.

In 1790 Nicholson took out his first patent for placing his moveable types on a cylinder ; but which he was never able practically to accomplish.

In 1811 König, after wandering for two years through Germany and Russia in fruitless search of patronage, found his way to London, and there, " by the aid of private individuals," says Dr. Gregory, " effected what could not be accomplished by the patronage of princes on the continent;"—this consisted in taking the impressions by means of rollers, from the types lying in a horizontal position.

In 1815 Cowper introduced his curved stereotype plates ; and in 1828 Applegarth so simplified König's machine as to dispense with not less than forty wheels, while he quadrupled the amount of production, and from 1100 sheets of the "Times" in an hour, increased the number to 4000.   To accomplish this, required nearly forty years ; still, the demand continuing to outstrip the supply, it became obvious that reciprocating motion must be abandoned sooner or later.   The great weight of the bed of type, moving at the rate of 225 feet in a minute, being suddenly stopped, while it involved a loss of nearly half the power, produced so great a shock and strain upon the machinery, that to attempt any increase on that velocity was impossible.

This fact Brunel had already established in the substitution of circular for reciprocating saws.   Applegarth at length succeeded in fixing the ordinary type to the outer surface of a revolving prism or drum, placed with its axis vertical—the column rules forming the wedges at the edges of the prism ; and so well were the types secured, that they remained uninfluenced by the centrifugal force created in working the machine.*

* The diameter of this drum was 5 feet 4 inches, and the circumference 16 feet 8 inches.

It was placed in the centre of the system and surrounded by *eight*

The result was, that whereas the reciprocating machine never delivered more than 6,000 impressions in an hour, the revolving machine accomplished with ease from 9,000 to 10,000.

This was justly considered a most valuable extension of the powers of the press, and so well satisfied was the " Times," that in a leading article, 29th December, 1848, it seemed to accept it as the *ne plus ultra*, and to abandon all hope of seeing the cylindrical press effective. " No art of packing," they then thought, " could make the type adhere to a cylinder revolving round a horizontal axis, and, therefore, aggravating centrifugal force."

American ingenuity has, notwithstanding, induced a material change of opinion and of practice, and the beautiful cylindrical machine of Messrs. Hoe and Co., with its six feeders, is enabled to double the boasted production of Applegarth, and to supply 18,000 sheets of this cosmical publication within the hour. By means, however, of a yet more recent improvement on Applegarth's machine, curved stereotype plates are obtained from the original type, thereby introducing an amount of economy in the preservation of the moveable type which more than realises the anticipations of Brunel.

In the invention of Major Beniowski, a farther increase of production is contemplated. The types, in

cylinders, their axes also vertical, upon which the paper was carried by tapes, and being connected with the central type drum by toothed wheels, the whole moved with a relative uniformity of motion; the central drum carrying the types, coming in contact successively with each of the eight drums carrying the paper. The types being eight times inked, would in one revolution leave impressions upon each of the eight sheets of paper on the cylinders.

place of being imposed on the outer surface of a cylinder, are attached to the interior surface. By this arrangement no amount of angular velocity can disturb the type bed ; and as his improvements embrace other economical appliances, the predicated results are from 20,000 to 40,000 impressions per hour.

Had Brunel's mind been earlier directed to the requirements of the printing press, it is not unreasonable to suppose that the same success would have attended his labours in that department of mechanical art which distinguished him in other departments, and that in place of nearly three-fourths of a century being required for the development of the press, and the successive application of so many minds, he would have given to the country a printing machinery, as he had a block machinery, within the limit of a few short years.

CHAPTER XI.

1821–1831.

MARRIAGE OF HIS ELDEST DAUGHTER, 1820 — MACHINE FOR
COPYING LETTERS, 1820 — PLANS FOR A BRIDGE AT ROUEN, 1830
— MUNICIPAL SYSTEM OF ENGLAND AND FRANCE CONTRASTED —
VISIT OF THE EMPEROR OF RUSSIA TO PORTSMOUTH — PLAN FOR
A TIMBER BRIDGE OF 880 FT. SPAN FOR ST. PETERSBURG, 1821
— AUTOCRATIC AND CONSTITUTIONAL GOVERNMENTS CONTRASTED.

BEFORE I continue the narrative of the public career
of Brunel, I may be permitted to refer to an event
in his home circle which proved not only a source of
social happiness and domestic consolation to him, but of
thankfulness and pride. This was the marriage of his
eldest daughter to Mr. Benjamin Hawes in 1820. To witness his son-in-law's enlightened sentiments recognised
by the liberal electors of the borough of Lambeth in
1832, and his public services subsequently rewarded
with the responsible post of Under Secretary to the
Colonies, was a subject of heart-felt congratulation to
Brunel. A few more years, and he would have had
the further gratification of seeing him Deputy Secretary at War, and ultimately permanent Under-Secretary of State for the War Department, with the
rank of Knight Commander of the Bath, an honour
conferred upon few connected with the civil service of
the country.

To return to Brunel's public labours. One of the

early productions of his inventive genius this year was a little machine for copying letters, and for which he secured a patent. The object of the machine was to enable writing to be transferred "by means of a damped medium, without the necessity for using wetting or drying books, and for other purposes."

The machine was called a *pocket copying press*, and for some years it appears to have been much sought after.*

But more important matter soon occupied his attention. From Rouen, the capital of his native province, he received an invitation to furnish designs for a bridge which should take the place of the old pontoon bridge, and permanently unite the town with the island of La Croix, formed by two branches of the Seine.

It will be remembered that it was at Rouen Brunel's mind received its first impression of British mechanical skill; at Rouen, too, his first deep feeling of attachment was awakened, which only terminated with his life; and for Rouen he always retained a strong interest, not to say affection. It was, therefore, with a grateful heart that he accepted the invitation of his friends and admirers. Two designs were speedily submitted, one to be executed in timber, the other in iron; but after various negotiations it was found, to his disappointment and the mortification of his friends, that unless he could obtain an appointment to the Government Corps of Engineers, the country must.be debarred the advantage of his services.

M. Quesnel writes (November, 1820): — "Quant à vous-même, je me permets à cause de l'intérêt que je

---

* See Appendix C for the description.

porte à votre famille et à vous-même de vous engager à
bien peser le parti que vous allez prendre ; et si en tra-
vaillant pour votre pays natal vous vous exposez à
perdre la confiance des étrangers, profitez de l'appui
que vous peut accorder notre ambassadeur pour ne
vous présenter en France qu'après vous être rattaché
au Corps des Ingénieurs des Ponts et Chaussées, soit
en qualité d'inspecteur, ou avec le rang d'ingénieur en
chef. Autrement je vous prédis que vous rencontrerez
tant d'obstacles que rien ne vous réussira."

MM. Dutens-Cordier and Dupin have each successively protested against the system so injuriously followed in France in the management of her public
works. Everything was undertaken by or through
the Government only. Private enterprise was thus
effectually checked, and the development of the resources of the country was for centuries paralysed.
"When the engineer-in-chief of a frontier department
draws up a plan of a road, bridge, canal, lock," &c.,
says Baron Dupin, "he must first send this plan to the
colonel of military engineers whose direction is in the
district where the proposed works must be executed.
But as there is no connection between the departmental
division and that of the military stations, the same
colonel of engineers is frequently obliged to discuss a
plan with the engineers of the bridges and roads of
two or three different departments. Now these officers,
who are perfectly independent of each other, judging
according to their own particular knowledge, scarcely
ever come to a mutual understanding. The matter is
accordingly referred to Paris, when arises the conflict
of different pretensions. As all work must have received the sanction of Government, and most works
were required to be undertaken exclusively by Govern-

ment, they were constantly postponed from the necessity of applying the funds to other more pressing demands, particularly in times of war, when money destined for the construction of a necessary bridge, or the repair of an important road, was allocated to the sustentation of the army and providing the muniments of war. While in England, by encouraging the spirit of enterprise in individuals, the Government wisely substitutes for the temporary expedient of taxation the permanent efforts of the public at large to administer to their own wants." As some evidence of the superiority of the one system over the other, we find that, "from 1790 to 1805, an interval which included four years of peace and eleven years of war, upwards of 1500 miles of canals were cut in England alone ;" whereas in France, not only were no great public works projected, but those which had been executed during the peaceful reigns of Louis XV. and Louis XVI., as well by the Government as by the provinces and cities, were entirely neglected, and those sources of wealth being exhausted, trade was paralysed and whole districts consigned to ruin.*

It is scarcely necessary to add that the difficulties suggested by M. Quesnel compelled Brunel to abandon the pleasing hope of serving his native city.

During the visit of "the crowned heads" to this country in 1814, Brunel had received from the Emperor of Russia the most distinguished marks of admiration, appreciation and esteem ; his Imperial Majesty having with his own hand placed a diamond ring on Brunel's finger as an earnest of the value which Russia would place on his services should the time ever arrive when she might command them.

* Dupin's Commercial Power of Great Britain.

The project of forming a permanent communication across the Neva at St. Petersburg recalled Brunel to the emperor's mind, and negotiations were entered into with him for the erection of a bridge over, or the construction of a tunnel under, the river. In 1818 Brunel had taken out a patent for "forming tunnels or drift-ways underground" by means of a gigantic boring machine; and in his memoir to Count Nesselrode, he gives a detailed account of the attempt which had been made under the Thames in 1808–9, and then describes a mode of construction calculated to overcome all those difficulties which had defeated the efforts of the first projectors. Into this discussion we shall not now enter, reserving our obser vations to the period when the "Thames Tunnel" was commenced.

The bridge, proposed to be constructed of timber, 880 feet span, "Exigerait à la vérité," he says, "une combinaison d'assemblages bien différente de celle que j'ai proposée dans d'autres constructions." And in a memoir which he presented through Count de Lieven to Count Nesselrode he discusses the manner of construction. "In order," he says, "to obtain the whole force of which timber is capable, it is only necessary to strengthen it laterally, so as to prevent all deflexion in the line of compression. As an indispensable condition, no mortise can be permitted. The principal pieces ought to preserve a uniform and proportionate dimension in their whole length.

"It is on this principle and on these data that I propose to construct a framing of carpentry capable of forming an arch 880 feet span." He then refers to the bridge at Tours as an example of the great weight sustained by piles.

"The arch," he continues, "which I propose may be considered as the segment of a large dome, the base of which would be 120 feet and the summit 30 feet. The horizontal thrust will not be greater in such an arch of 880 feet than in a stone arch of 60 feet ; and as there will be no support required other than from the abutment, this arrangement will be found highly economical in structures of such length. The dome may be covered with copper or zinc.

" It is easy to see that each half of this arch cannot bend ; and that the arch itself cannot yield to the weight which it is intended to support, so long as the abutments resist the thrust. In this respect there is no difference between an arch of wood and one of stone, so long as the weight and the angle of it remain the same.

" It is easy nevertheless," he goes on to say, " to satisfy oneself as to the effect of compression on beams a foot square, where the versed sine is the seventh part of the span of the arch. Fir timber being the best adapted to the construction of an arch of such dimensions, I base my calculations on the ratio of the resistance of such timber, both to compression and tension in the length of the fibres.

" Timber is the principal material used in almost all constructions. One observes everywhere timber forming the foundation to stone. The most solid and colossal monuments, abutments, piers of bridges, rest on the heads of a given number of piles which are required not only to support the weight, but to resist also a lateral pressure, such as abutments of bridges. If, under such circumstances, the timber were to yield one inch after the bridge had taken its set, the whole structure would be fractured, and the consequences so serious as to render the structure a failure."

The accompanying sketch exhibits the plan and elevation of half the arch, and the suspended roadway broken in the centre for the passage of vessels. It also exhibits the importance of the ties B B, employed to give the necessary stability, and to prevent all movement between the abutment and the centre A.

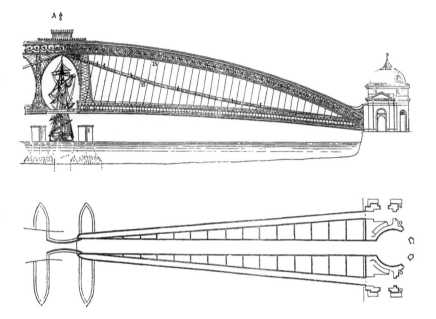

In describing the means by which he proposed to erect the bridge under the peculiar circumstances of a river subject to be yearly frozen, and which would present, when the ice broke up, further difficulties, he says : " To resort to the ordinary methods would be hazardous." He then recommends " that the bridge should be constructed in some convenient locality, sheltered from frost, and when completed, and the river free, floated into its place. Four pontoons of dimensions proportioned to the magnitude of the struc-

M

ture," he says, "would easily form the platform; for,
independent of the reasons which suggest this expedient, it will be found really less costly and more
certain than any other."

Why this magnificent, bold, and original project was
not adopted will be seen from the following despatch,
addressed by Count Nesselrode to the Russian am-
bassador, Count de Lieven, $\frac{30 \text{ April,}}{12 \text{ May,}}$ 1821, who says :—

"L'Empereur, qui estime depuis longtemps et ap-
précie les profondes connaissances de l'ingénieur Brunel,
a lu avec un vif intérêt la dépêche de votre Excellence,
No. 45. L'attention de sa Majesté Impériale s'est
particulièrement fixée sur les plans que M. Brunel propose pour l'établissement d'une communication entre
les deux rives de la Néva. L'Empereur les a examinés
lui-même, et ne peut qu'être vivement obligé à leur
savant auteur de les avoir portés à sa connaissance.
Sa Majesté Impériale vous charge, néanmoins, Monsieur
le Comte, à déclarer franchement à M. Brunel que les
circonstances présentes ne favorisent pas l'exécution de
ses projets; et que forcé à des dépenses imprévus et
très considérables, le trésor impérial ne peut subvenir
aux frais d'une aussi coûteuse entreprise. Comme elle
ne saurait l'accomplir en Russie au moyen des sou-
scriptions usitées en Angleterre, et que la ville de Pétersbourg ne possède pas de fonds municipal que l'on puisse
spécialement affecter à de semblables travaux, l'établissement d'une communication permanente entre les
deux rives de la Néva doit nécessairement être ajourné.

"Sa Majesté Impériale lui laisse en conséquence
la libre option — ou de remettre son voyage à une autre
époque, ou de se rendre à Pétersbourg uniquement afin
d'y parcourir les établissemens publics qui pourront pro-

fiter de ses lumières et de sa grande expérience. Si M.
Brunel prenait ce dernier parti, comme la présence d'un
homme de son mérite est toujours désirable, et que selon
toute probabilité nous retirerions plus d'un avantage de
sa visite dans nos chantiers, ainsi que des conseils qu'il
donnerait sur l'usage de nos bois de construction, vous
voudrez bien lui annoncer, Monsieur le Comte, qu'il
recevrait en Russie l'accueil dû aux talens qui le dis-
tinguent et aux intentions même qui l'ont porté à nous
offrir ses services.

"L'Empereur désirerait en ce cas que M. Brunel ne se
rendit à Pétersbourg qu'au moment où sa Majesté sera
de retour dans sa capitale.

"Il est entendu que l'Empereur l'indemniserait alors
des frais de son déplacement, et vous nous indiqueriez
d'avance la somme qu'il conviendrait de lui envoyer à
cet effet.

"Recevez, Monsieur le Comte, l'assurance de ma haute
considération.

(Signé)   "NESSELRODE."

It is curious to mark the condition to which these
two great nations—Russia and France—were reduced
by the defects in their municipal systems.

Local public works, however desirable and important, even to the general welfare of the countries, could
not be executed, because the principle of centralisation
had robbed the people of all control over their own
affairs.

M. Dutens, writing in 1819, says:—

"In truth, this disposition on the part of the English
Government, to abandon the execution of canals, &c.,
to companies, is less the effect of a theory than the consequence of an administrative system, which naturally

flows from the most important revolution which ever affected the constitution of a country, and of which the extended influence is perceptible, even in the smallest ramifications of the administration. In becoming altogether municipal, the immediate power of the Government is evaded in a manner nowhere else accomplished. In point of fact, few problems of administration in England can be solved without immediate reference to the period and to the spirit of the revolution of 1215.

" The nobles having called the people to their aid, and having united with them against the tyranny of John Lackland, produced a combination against the royal authority, which extended, under certain limits, the domain of municipal administration, and, in consequence, tended to embrace successively all kinds of public works, the necessity for which affected only the local interests, and the knowledge of which could scarcely be brought to the cognisance of the sovereign. In France, on the contrary, the revolution, accomplished nearly at the same time, produced a diametrically opposite effect.

" The kings, in contemplating the condition of the people, thought less of enfranchising them than of weakening the nobles.

" It was not then the nation which wished to become great — it was the monarch alone that desired to become strong and powerful." *

* Mémoire sur les Travaux Publics de l'Angleterre.

CHAPTER XII.

1814–1821.

WHILE the negotiations with the Emperor of Russia were still in progress, the sawing establishment at Battersea had fallen into inextricable confusion.

We have seen that in 1808, Brunel had, in conjunction with the Messrs. Farthing, erected a sawing establishment at Battersea, which promised to secure an ample fortune to the inventor. So effectively indeed did it at one period fulfil the anticipations of the proprietor, that I find 4726l. 5s. 9d. was actually received from builders, cabinet-makers, &c., in five months, or at the rate of 11,343l. 2s. per annum.

The expenses are set down at 3200l., leaving, therefore, a net profit of 8143l.

Other partners had subsequently taken the places of Messrs. Farthing, and to them Brunel continued implicitly to confide his pecuniary interests; but, notwithstanding the undoubted success of all the mechanical arrangements, difficulties soon began to be felt in the financial department, and which were brought painfully to light by an event as unlooked for as it was calami-

tous. On the night of August 30, 1814, a fire broke
out which, in two hours, nearly destroyed an establishment which had been valued at 24,000*l*., and which had
cost many hours of anxiety and self-denial. At the
time that the fire was discovered, Brunel was at Chatham ; and, to add to the difficulties, the principal fire
engines were unfortunately engaged in the City, where
a more disastrous conflagration was still raging, and
which had destroyed, at Bankside, the mustard and oil-
mills of Messrs. Wardle and Jones, the corn-stores of
Mr. Hammocks and Messrs. Resden and Ayres — the
hop warehouses of Messrs. Clarke and Myer and Mr.
Evans, the iron foundry of Messrs. Ball and Jones, and
a dyeing establishment of Messrs. Thell and Steele.
Nor was this all. The oil, finding its way to the river,
spread itself over a large surface of water, involving
coal barges with their freight in the conflagration.
Almost all the fire-engines of the metropolis were engaged, and all that could be obtained for Battersea were
two engines of Mr. Noble, and one from Chelsea. To
complete the disappointment, and destroy all hope of
rescuing the property from destruction, the tide was
out, and scarcely any water could be obtained to supply
the engines. " Thus, in two hours," says the writer in
the "Gentleman's Magazine" (vol. lxxxiv. pt. ii.), "these
most valuable machines, which, in point of execution
and perfection, exceeded everything we know, and
which had been visited by some of the most illustrious
characters in Europe, presented the awful sight of a
heap of fragments ; and the fruits of six years of exertion and ingenuity, attended with an expense of above
20,000*l*., were destroyed."

The right wing of the building and the steam engine
were, however, saved. No time was lost by Brunel in

vain regrets. With that wonderful elasticity of spirit, and confidence in his own resources, he at once sought for means to repair his great loss. This was not to be so readily obtained, for, to his mortification and amazement, he discovered that out of a capital, which on the previous October amounted to 10,000*l*, there remained on the 15th June, when the foundations of the new work were commenced, only 865*l*. to the credit of the firm, in their banker's books. " Still relying," says Brunel, "upon the pecuniary aid I could bring in, I pushed the enterprise without interruption. Availing myself of the experience I had acquired, I directed my attention to all the improvements that could be introduced consistent with our scanty means."

True, the machinery was replaced, and the mechanical value of the concern was restored, if not increased ; but no amount of ingenuity could call back the balance in his banker's books.

In his difficulty Brunel sought the advice of a Mr. Sansom, a banker in the city, a personal friend, and, to judge from his correspondence, a gentleman of strong understanding and business habits. He at once proceeded to investigate the affairs of the partnership, and from his communications may be gathered the ruinous position into which Brunel had been brought by a misplaced confidence.

Not that Brunel was incapable of entering into calculations the most detailed, nor that he was unacquainted with the requirements of trade — quite otherwise. His correspondence shows how entirely he appreciated all the conditions by which commercial success can alone be commanded. It was his misfortune, rather than his fault, that those to whom he confided his pecuniary interests failed to fulfil the duties which they had accepted; and no amount of intellectual resource on his part could shield him from the consequences of that fatal connection.

Mr. Sansom very soon formed his opinion of the business management, and he thus writes to Brunel, November 1816 : — " The more I investigate, the more I feel it necessary for your safety that you should sift to the bottom every cash transaction ;" and, as one striking cause for the great depreciation of the credit of the establishment, he mentions, " the single circumstance of the allowance of 20 per cent. discount upon the work done at Battersea ! " " What must be the interpretation of the party to whom it is allowed ? " he writes, —" one of two things most certainly—either you charge an extravagant profit, and can therefore allow it, and perhaps ought to allow more, — or that you must have money, *coûte que coûte*."

On the 23rd of the same month he again writes, " You might make the gross produce of your mills 8000*l*. or 10,000*l*. per annum. I think the latter sum even fairly within your reach."

Unfortunately Brunel was induced to permit the shoe operations to be mixed up with the saw establishment, the result of which was a complication of accounts that induced his friend Mr. Sansom to write, January 7th, 1817 : —

" It was a most extraordinary jumble which you certainly have not understood, and I should have wondered if you had. I should hardly have been more surprised than I am if one of your saws had walked to town."

This complication necessarily introduced a variety of legal questions, with all the perplexing and expensive references to legal opinions, which drew from Mr.

Sansom, in a communication, January 1817, the following observation, semi-jocular, semi-serious : —

" If you have ever been ill in your life, and have
depended upon medical advice, fall down on your
knees and bless God that you had fewer doctors than
you have had lawyers about you. If that had not been
the case, you might have been making saw-mills on the
other side of Styx, or inventing a steam-boat for old
Charon."

It would serve little purpose to record the progress
of a series of painful investigations, so actively prosecuted by Mr. Sansom, except in so far as they were
found to have influenced the character or directed the
conduct of Brunel. Suffice it to say, that a change was
effected in the partnership, and that a friend of Mr.
Sansom ultimately became connected with the concern;
though, unfortunately, without producing any ultimate
benefit to Brunel. Of the negotiations which preceded
this change, Mr. Sansom thus speaks : —

" Your conduct has, from the commencement of our
treaty, been in the highest degree honourable, liberal,
and friendly ; and, though I am very anxious for my
friend, I can sometimes hardly help regretting that I
could not *myself* embark with you. Your business at
Battersea, I can very clearly see, is to be made a very
lucrative one, and, if I were your partner, I would
answer for showing you a very different balance sheet
for the year 1818."

Notwithstanding Mr. Sansom's hopeful opinion, difficulties continued to accumulate, and against which
Brunel was left single-handed to contend.

In a letter to Mr. Sansom, he says :—" It is no less
painful than it is discouraging to reflect that, independent of the sum I have lost in my late connection, part

of that sum has been brought by me for the purpose of effecting the restoration of our present saw-mills, in consequence of which the parties interested with me have been placed in a higher situation than their original share entitled them to. This consideration appears to be entirely lost sight of. Alone I am to support the whole burden. *I am called upon to pay for what belongs to me*, in order to balance the account of others." The large losses which he had incurred in the erection of the shoe machinery incapacitated him from meeting the demands to which he became ultimately liable, owing to the withdrawal of the principal acting member of the firm from the country. The failure of the bankers, Messrs. Sykes and Co., brought matters to a crisis. Bills to a considerable amount were dishonoured. "Hungry ruin had him in the wind;" an execution came upon his house; he was arrested (14th May, 1821), and in the confinement of a prison he was made to feel that to secure commercial success, the intelligence and integrity of agents are as important as the conceptions and inspirations of genius.

That the shock was deeply felt may be well believed, nor would it have been matter of surprise if, like the amiable and learned Floyer Sydenham, Brunel had also sunk under the pressure of his calamity *; it is, therefore, worthy of remark, that throughout a variety of negotiations, tiresome delays, and untoward incidents, the cheerful and hopeful spirit of Brunel still preserved to him the most advantageous exercise of his faculties. The shortcomings of others produced in his mind no bitterness, nor did they relax for an hour his efforts to

---

* It is recorded that Sydenham, after he had been arrested for a small debt, never again spoke.

relieve those with whom he was connected from the consequences of their own actions; whether, by a liberal application of his own pecuniary resources, in the sacrifice of his former earnings, or in the untiring exercise of his inventive faculty. Here, at the very period of his arrest, we find him, in fact, in correspondence with the Russian Government, and a prospect opened to him of permanent and highly remunerative employment. Surrounded as he was by difficulties, oppressed as he was with anxieties, had he felt less regard for social, political and religious liberty, he would scarcely have resisted the temptation which Alexander so freely offered. That the struggle was severe we may well believe. To be robbed in the land of freedom of that personal liberty, for which he was willing to sacrifice so much, was, indeed, a grievous, almost an overwhelming sorrow.

No sooner was it known to his numerous and influential friends, and to those who could estimate the value of his unprecedented services, that he was no longer in a condition to contend against the pecuniary difficulties which beset him, than every effort was made to obtain the means of releasing him from his painful position.

Amongst those who took the most active part in urging his claims upon the Government, at the head of which was his Grace the Duke of Wellington, were Earl Spencer, Dr. Wollaston, Sir Herbert Taylor, Sir Edw. Codrington, Mr. W. Smith, M.P., Mr. Arbuthnot, and Mr. Bandinel of the Foreign Office. We have seen how strong the temptations were to induce Brunel to take service under the Imperial Government of Russia, and now he writes to Mr. Bandinel from the King's Bench, June 1821 :—

" If I had been indifferent on the score of attachment for this country, I should long ago have accepted
the flattering offers I have had of employment abroad,
and, indeed, it is not without just ground if I had so
done. I beg you will consider that for the last four
years I have had nothing to do for Government, consequently no pay — no tie whatever,— and that I have
been subjected to many vexations.

" Instead of availing myself of the opportunity of
obtaining employment abroad, and I may say honour
too, I have endeavoured, by all means in my power,
and by every exertion possible, both on my part and
through my distinguished friends, to obtain such relief
as the peculiar circumstances of my case had entitled
me to. From 1818, the period of the Congress of Aix,
when my plans were before his Imperial Majesty, I
have delayed sending my proposition and terms for establishing a communication across the Neva. It is only
in December last that I sent plans and propositions for
either a tunnel or a bridge across the river. The latter
plan, which is a most extraordinary plan, and on a most
enlarged scale, was honoured by the peculiar notice of
his Imperial Majesty the Emperor." He further says
that he had " prepared plans relative to the economy of
timber in dockyards, and to the mode of building ships,
and that, although peculiarly calculated for Russia,
they would be attended with very great saving and
advantage in this country." And he closes his communication by stating, that " if I see honourable and permanent employment here, you may be assured that I
shall not be wanting in zeal, but shall devote my future
services and talents for the benefit of this country."

Strenuous efforts continued to be made to induce the
Government to relieve Brunel from his pecuniary diffi-

culties. Mr. Bandinel, who took upon himself to conduct the negotiations, writes to Mr. Benjamin Hawes.

" My dear Sir,

" Mr. Arbuthnot has told me that it is the intention of Government to do something immediately with a view to relieve Mr. Brunel from his present difficulties. He added that a report was current, that Mr. Brunel would, on being released, go to Russia, and that, if such were to be the case, Government would not relieve him ; for the step they now take is more in liberality than absolute justice, and they have a right to hope for the benefit of his future services. I answered than I had, in the first instance, made it my business to come to a distinct understanding on this point with Mr. Brunel, who had told me that it was not his intention to go to Russia if he could get employment here ; that upon his relief from his difficulties, he should apply to Government for employment. If they would not give an immediate answer, he should still wait, and if in the mean time a distinct offer came from Russia, he then should go to the Government, and say : I applied to you for employment — you gave no answer — here is an offer from Russia. I must starve, or get employment here, or go to Russia. You cannot expect me to starve ; will you give me employment, or shall I go to Russia ? and *then only* should he, upon that further declining to give him employment here, go to Russia.

" June 18th, 1821."

" Fleet footed is the approach of woe,
But with a lingering step and slow
Its form departs."

Wearied at length by these protracted negotiations, — unable to cheat himself longer into forgetfulness of the present, worn out, humbled, distracted, the agony of his free spirit found expression in a letter to his early and faithful friend, Lord Spencer : — " My affectionate wife and myself are sinking under it," he writes. " We have neither rest by day nor night. Were my enemies at work to effect the ruin of mind and body, they could not do so more effectually."

" Thus many a sad to-morrow came and went," and he that had enriched hundreds by the exercise of the most honoured of the human faculties, was left for months to mourn the hardness of his fate :—or, rather, having permitted himself to confide too implicitly in others he was condemned to suffer the penalty which a transgression of a great moral law so sternly claims, and from the operation of which no plea of ignorance, no exaltation of special faculties, can exempt the transgressor.

Ultimately, in consideration of the distinguished services which Brunel had rendered the country — more particularly in relation to the block machinery,— it was determined that relief should be given to him by the Government, and accordingly a sum of 5000*l.* was placed by the Treasury in the hands of Mr. Bandinel to discharge the liabilities of Brunel.

To the aid which, above all others, his Grace the Duke of Wellington had afforded during the progress of the negotiations with Government, Brunel felt deeply grateful, and on the 10th of August, 1821, he endeavoured to convey the sense of his obligation in the following communication : —

" My Lord Duke,

" I was very much disappointed at being deprived, through your Grace's absence from England, of the opportunity of returning in person the thanks I owe to your Grace, for the favour shown to me by the British Government in the adjustment of my affairs, and in the consequent liberation from my confinement.

" Sensible as I may be at the happy termination, I cannot find expressions to say how much I feel at the peculiarly fortunate circumstance of having been deemed worthy of the notice and patronage of the most distinguished character of the age — a circumstance not only most flattering and most honourable, but which has materially contributed in softening the gloom which so distressing a reverse would otherwise have left on my mind. The only way by which I could make a suitable acknowledgment to your Grace was in employing my time in preparing plans for the service of the British Government."

Any latent disposition which Brunel might have ever entertained to connect himself with Russia was now entirely removed, when he found himself once more a free citizen under a constitutional government.

CHAPTER XIII.

1821–1826.

SUPERVISION OF THE WORKS AT CHATHAM, 1821 — SAW-MILLS
FOR TRINIDAD, 1821, AND BERBICE, 1824 — SUSPENSION BRIDGES
FOR THE ISLE OF BOURBON, 1821–1822 — DIFFICULTIES WITH
CONTRACTORS. — MARINE STEAM ENGINE, 1822 — IMPROVEMENTS
OF PADDLE-WHEEL, 1823 — LIVERPOOL DOCKS, 1823 — CARBONIC
ACID GAS ENGINE, 1824 — FAILURE — OPINION ON TRANSATLANTIC
STEAM NAVIGATION — ISTHMUS OF PANAMA, 1824 — SOUTH LON-
DON DOCKS, 1824 — RAILWAYS IN FRANCE — CHESTER BRIDGE,
1825 — RUBBLE BUILDING, 1824 — FOWEY AND PADSTOW CANAL,
1825 — BRIDGE AT TOTNESS, 1825 — VIGO BAY, 1825 — LIVER-
POOL FLOATING PIER, 1826.

RESTORED to his home, with a heart grateful for the
liberality which had been accorded him, we find
him resuming the general supervision of the works at
Chatham, and in designing, under the authority of
Government, saw-mills for the island of Trinidad—
that "Indian Paradise," as it has been called, and which
possesses so much of historic interest.

Gladly would Brunel have renewed his early recollections with that beautiful country, had circumstances
permitted him to have superintended his own works;
but more important labour was at hand. The projection and estimates for Trinidad were followed in
1824 by similar ones for Berbice in British Guiana.
This country abounds with forests of every variety of
timber; some growing to the height of eighty to one

hundred feet without a branch ; the valuable *Ferole* or *satin-wood;* the *Licaria* or *rose-wood;* two kinds of *Icica;* the *berk-back,* the *mahogany* and the *cuppy ;* while the *Andera bulata* and *Onatapa,* though susceptible of the highest polish, could only be cut with the greatest difficulty from their extreme hardness. In short, in no part of the world does nature put forth her power of vegetation so largely, as on the banks of the Demerara, the Berbice, the Courentin, the Canje, the Sinamari and the Arauari. There may be seen leaves — those of the Troolies—from twenty to thirty feet in length, and from two to three feet in breadth, supported on stems three inches in circumference, forming a protection against the most violent tropical rains, and, as a roof, found capable of affording protection for many years. Here then was a locality admirably adapted for the application of the saw, and Brunel's early acquaintance with the character and properties of the vegetation of that country enabled him to select such machinery as was best adapted to the circumstances.

Singular to say, in the communications which passed between General Mann on the part of the Ordnance Department, relative to the number and extent of the works to be executed by Brunel, and the instruments to be employed, so little importance was attached to the value of the circular saw, that Brunel deemed it right to recall attention to that important instrument.

When furnishing his reduced estimate, he says: " Now with regard to circular saws, I have to remark that as their utility and advantage have been entirely lost sight of, it is the more necessary in the present statement to notice it, as it is a very important item in the return that may be expected from the mill on its reduced scale.

" The employment of circular saws is found to answer

N

peculiarly well for the conversion of a great variety of articles, which there is generally a large consumption of: such as lathing for roofs, for palings, gates, rails, posts, rods, helves, &c., and all kinds of rabbeting.

" It is found that all work done by the aid of circular saws is paid by piece-work *one-ninth* part only what is paid when done by the hand."

The cost of the machinery for one mill and erection, was estimated at £7500.

The yearly expense of working	.	.	£1095
The interest on outlay @ 8 per cent	.	.	600
			——— 1695 0 0
While the lowest rate of produce was valued at			4143 15 0
Leaving a clear yearly profit of	.	.	£2448 15 0

That this result could scarcely have been considered speculative may be fairly concluded, when a comparison is instituted between the work actually accomplished at Chatham this very year by means of the circular saw and the same work executed by hand.

Thickness of Wood.	Cost by Hand (Oak) Per 100 feet.	Cost by Hand (Fir) Per 100 feet.	Cost by Circular Saw.
Inches.	Pence.	Pence.	Pence.
2	$11\frac{1}{2}$	9	$1\frac{1}{4}$
3	16	13	2
4	21	$16\frac{1}{2}$	$2\frac{1}{2}$
5	26	$20\frac{1}{2}$	3
6	31	$24\frac{1}{2}$	$3\frac{1}{2}$
7	36	$28\frac{1}{2}$	4
8	$40\frac{1}{2}$	$32\frac{1}{2}$	$4\frac{1}{2}$
9	$45\frac{1}{2}$	$36\frac{1}{2}$	5
10	50	40	6

A commission was issued this year by the French Government to obtain designs for suspension bridges to be erected on the island of Bourbon, and to Brunel the commissioners applied. This opportunity of exercising his talents in the service of the country of his birth, although devoid of the special interest attached to Rouen, was still to Brunel a source of honourable pride, and he entered upon his task with that zeal and originality by which all his designs were characterised. A new field for the development of his constructive talent was now opened to him ; for, although in that favoured island there may be gathered examples of the vegetable riches of the whole eastern world in its *jardin de l'état*, yet, what was of more importance to Brunel, it was subject to hurricanes moving with a velocity of 120 to 150 feet in a second, or from 82 to 102 miles in the hour, to resist which his ingenuity and skill would be abundantly taxed. Without reference to authority, but guided only by a clear conception of the conditions to be fulfilled, we find Brunel presenting at once a model of mechanical completeness. Not only was provision made to compensate for the expansion of the metal, and therefore for the elongation of the suspension rods, but to resist the destructive force of those fearful hurricanes to which we have referred.

It is true the span was not great, little exceeding 122 feet in the clear ; yet the design will be found to bear a favourable comparison with the more recent and more extensive works of a similar character in this country, both as regards beauty of form and efficiency of construction.

Each bridge was divided into two roadways, suspended by six chains $1\frac{3}{8}$ inches diameter, with suspension rods $1\frac{3}{4}$ inches diameter. The timber used in the platform was of teak ; the proof to which the bridges were subjected, was a weight of 25 tons, laid on 80 feet of the centre of each catenary ; or 70 lbs. per square foot, and there permitted to remain for 41 hours.

The peculiarity of the construction consisted in the introduction of four chains curving upwards and sideways, fastened at the ends to the abutments and centre pier. By this arrangement the roadway was not only suspended, but confined between two systems of arcs, which formed a general bracing capable of being tightened at pleasure (see plate).

On the 20th of October, 1821, the French Government accepted the designs and specification of Brunel; but it was not until the 19th July, 1822, that the agreement was ratified, and a contract for 7000*l.* concluded between Brunel and the " *Ministre Secrétaire d'État de la Marine et des Colonies,*" the " *Marquis de Clermont-Tonnerre ;*" the work to be executed in four months, and the money to be paid by four instalments through *Messrs. Outrequin and Jauge.* The Bowling Iron Works of Messrs. Sturge at Bradford seem to have been first selected for various experiments, but the construction of the bridges was finally, and, as events too painfully showed, unfortunately for Brunel, entrusted to another firm. The same fatality which had already marred Brunel's commercial prosperity was found still to cling to him.

From various causes, the completion of the work was most unjustifiably delayed; compelling Brunel, in December 1822, to enter a formal protest against the breach of contract, " thereby," he says, " subjecting me to great loss and injury, and to have the bridges, when completed, rejected ; " and as the new year opened, the extreme cold which prevailed operated most prejudicially on the progress of the work. Brunel records in his journal, 15th January, 1823, " the cold so intense that cast iron 15 × 10 inches was broken." On the 29th April, 1823, the bridges having been set up by

BOURBON BRIDGE.

the contractors, they were inspected by M. Sganzin, *Inspecteur-général des Travaux Maritimes*, who accepted them on the part of the French Government. This important preliminary accomplished, it might be naturally concluded that the bridges would be at once taken to pieces, and despatched to London for shipment in accordance with the terms of agreement. But no; charges for extra work were preferred by the contractors, to the amount of upwards of 500*l.* ; and they refused to forward the work to London unless their demand was complied with. Brunel resisted ; and a long correspondence ensued, the lawyers alone deriving any advantage. At length, in July, the bridges were taken to pieces under the supervision of Brunel's intelligent and incorruptible superintendent, Thomas Mathews, who continued to furnish detailed reports of the progress of the work, together with evidence of the various efforts made to elude his watchful eye, and evade the requirements of the contract,—weighing the ironwork when his (Mathews) back was turned, or when work had stopped for the night.

"I told 'em," says honest Mathews, "of all the jobs I ever had been at, I never saw such goings on as there."

At length, by the middle of August, the bridges packed in cases arrived in London ; but from the doubts which had been thrown on the proceedings at Milton, Brunel felt it his duty to have the cases examined, and the iron reweighed. Writing to the London agents of the contractors, he says : "This leaves no alternative but that of opening all the casks and boxes, and perhaps all other parcels, and of inspecting, taking an account, and of weighing the contents. Relying on the characters of those gen-

tlemen (the contractors), I did not think it necessary
to scrutinise their accounts; but on examining them,
I discover such deficiency in some of the component
parts of their bridges, as to render that inspection
necessary." There was found, in fact, a deficiency
of 800 feet in the length of the under-chains, while
the flat links wanted 200 of their proper number.
Anxious to fulfil the anticipations of a country which
had never ceased to regret his expatriation, and which
had now availed itself of his services, although he
remained unconnected with its privileges, Brunel felt
grieved as well as disappointed; and as a considerable
sum remained to be received when the bridges were
delivered to the Government agent, he was also put to
a considerable amount of pecuniary inconvenience.

At length M. Sganzin writes from Paris, August 31st,
1823 :—

"J'ai appris par son Excellence, à qui M. le Baron Séguier l'a mandé, que les ponts pour Bourbon étaient enfin
arrivés sur le quai à Londres, et par conséquent prêts
à être embarqués. J'ai remis au Bureau des Colonies
tous les documents nécessaires pour que l'ingénieur de
Bourbon puisse procéder au montage des ponts immé-
diatement à leur arrivée.

"En vous remerciant de nouveau pour les gravures
des ponts que vous avez bien voulu m'envoyer et que
j'ai reçues, je dois vous prier de m'envoyer la note de
vos déboursés pour ces acquisitions de gravures, ainsi
que celle des frais pour la copie des dessins en grand de
vos ponts que dans le temps je vous ai prié de me
faire faire, et qui serviront à l'exécution du modèle que
son Excellence désire avoir dans son cabinet. Aussitôt
que cette note me sera parvenue, les sommes vous seront

remises par l'un des correspondants de M. Jauge, qui m'a promis de me rendre ce service.

" Agréez, je vous prie, Monsieur et excellent ami, l'assurance des sentiments de l'inviolable attachement que je vous ai voué.

"J. SGANZIN."

The bridges were not shipped until the end of October. In the operation of shipping further defects were discovered, and the Messrs. Bourdieu and Co., in reporting to Baron Séguier, the French Consul-general, state that "four of the cast-iron bearings have been broken, and all the persons concerned have remarked upon the badness of their quality." At length, on the 29th November, the bridges left the river for their destination, and the overcharge made by the contractors having been withdrawn, a final settlement was effected, and Brunel's mind was at rest.

In 1822, Brunel took out a patent for " Improvements in Marine Steam Engines."

The objects were :—

1st. So to combine the power of two engines, that they might be applied in the most direct way for the production of rotatory motion.

2nd. To moderate and regulate the movements of those engines, and to produce an equable action.

3rd. To supply a mode of condensing the steam.

4th. To make certain adaptations to the boilers.

For a description of the improvements see Appendix D.

The year 1823 seems to have been singularly fruitful to Brunel in mechanical and commercial projects.

Early in the year I find him engaged upon improvements in the paddle-wheel for steam-vessels.

Perhaps there is no piece of simple mechanism upon which so many opinions have been offered, or in which so many efforts have been made to introduce improvements, as the paddle-wheel.

We have seen the parallel, the radiating, the cycloidal, the folding, the divided, and the reefing, each supported by its own theory; but each in its turn failing to satisfy all the necessary conditions in practice, and each in its turn rejected for the "old Jonathan Hull's" despised but still efficient instrument of propulsion. That many of these ingenious but abortive efforts have resulted from a misconception of the circumstances under which the paddles act, seems obvious. The same loss of power found to obtain at the commencement of the paddles' motion, was erroneously attributed to it when in rapid action, the effect of the contemporaneous motion of boat and paddle having been overlooked. It was, however, found that this resulted in a curve, which really fulfilled in the most ample manner all these conditions that the ingenuity, labour and expense of years had in vain sought to obtain; still there remained the great desideratum of applying a wheel which should act equally well when the vessel is deeply laden, and when lightened of her cargo, and to this Brunel addressed himself.

For the docks at Liverpool Brunel supplied this year some beautiful plans for swing-bridges, and for the Serpentine river in London a very elegant design for a suspension bridge.

He was also employed in preparing designs for improvements in the tread-mill, and for boring of cannon for the Government of the Netherlands. In the latter, certain beautiful mechanical improvements were introduced, while the beds, in place of being parallel to one another, were made concentric. This arrangement permitted the crane to be placed in the centre, and the jib to swing all round the standard.

No sooner had Sir Humphry Davy communicated to the Royal Society the result of Faraday's experiments upon the liquefaction of gases, in the spring of this year (1823), than Brunel sought to give the discovery a practical application.

Faraday showed, in a manner as simple as it was beautiful, that by the application of cold, certain gases could be liquefied. Into a bent tube, having one arm considerably shorter than the other, and hermetically sealed, he poured through a small funnel, down the longer arm, as much strong sulphuric acid as would nearly fill the shorter. A piece of platinum foil was then crumpled up and pushed in, and upon the foil was dropped fragments of carbonate of ammonia, until the longer tube was nearly filled ; the foil effectually impeding chemical action until the longer end of the tube was sealed.

This accomplished, the sulphuric acid was brought into contact with the ammoniacal carbonate, when the gas was evolved, and the long end of the tube being immersed in a freezing mixture, an extremely light, limpid and colourless fluid was produced. It was subsequently shown that the vapour of this fluid exerted at 32°, or the freezing-point of Fahrenheit, a pressure of thirty atmospheres or 450 pounds on the square inch, and that by the application of hot water only at 112° a pressure of sixty-five atmospheres was obtained. To control and utilise a power of this magnitude called for a considerable amount of metallurgic as well as mechanical skill. The first difficulty which presented itself to Brunel was to find such a combination of metal as would resist the pressure when the temperature was sufficiently elevated to yield an available motive power.

Through cast iron the gas passed as water through a sponge, though only subjected to a temperature of hot water at 112°. After a lengthened series of costly experiments, extended over two years, a sort of gun metal was obtained, which was found capable of resisting the subtle pressure of the gas, and with it a machine was constructed. This consisted of two sets of strong metal cylindrical vessels communicating with a working cylinder, one at the top, the other at the bottom.

The gas being obtained from carbonate of ammonia and sulphuric acid under a gasometer, was condensed by means of a condensing pump into the vessels, called "receivers," each containing a number of thin metal tubes. The receivers communicated with other metal cylinders called "expansion vessels," and these directly with the working cylinder. Hot water being passed into the thin tubes in one condenser at 120°, and at the same time cold water into the tubes in the other, the power developed in the one was found to be equal to about ninety atmospheres, while that in the other did not exceed fifty-five, leaving, therefore, a difference of thirty-five atmospheres as the available force to be applied alternately to the top and bottom of the piston in the working cylinder, as the fluid in the condensers was alternately heated and cooled.

Patents were secured for Great Britain, Ireland and France. Mr. Lilley and Mr. Howard sought to oppose the grant. The former had adopted the same principle in the generation of the gas; but when the invention was referred to the Attorney-General it was found that the mechanical arrangements differed altogether from those of Brunel.

The benefits anticipated from the gas machine strongly excited the hopes of the scientific as well as the commercial world. Brunel extended his patent to Holland and the Netherlands. The attention of the Government was aroused, and the Admiralty ventured to make an advance of 200*l.* to aid Brunel in obtaining a practical result. From France and Belgium orders for machines were received, but all in vain; and it only remains for me to add that, notwithstanding the mechanical difficulties had one after another been met and overcome; that the condensation of the gas had been effected under pressures extending to *three hundred atmospheres;* that the joints of the tubes were made capable of sustaining a pressure of *fifteen hundred pounds* on the square inch; that upwards of 15,000*l.* had been expended in the necessary experiments, and nearly fifteen years devoted to the solution of the problems,—yet the conclusion to which Brunel was compelled ultimately to submit was, " that the effect of any given amount of caloric on gaseous bodies was not greater than that produced by the expansion of water into steam; " and that, therefore, " the practical application of condensed gases, including common air, was not so advantageous as that derived from the expansive force of steam." * Thus this beautiful

* See an interesting discussion at the Institution of Civil Engineers, upon Sir George Cayley's hot-air engine, February 1850. Mr. Isambard K. Brunel, whose opinion I have quoted, had superintended all the important experiments for his father in 1824–1825.

theory, which had given so much promise, and which had been hailed as the harbinger of a new era in practical mechanics, was found incapable of realising those economic conditions by which alone it could be rendered commercially valuable.

I may here state that while Brunel was engaged upon the gas experiments, he was induced to offer an opinion upon the extension of steam navigation, so curious, so unexpected, and apparently so little capable of being reconciled with the natural tendencies of his mind, and with those improvements going on around him to which he was himself largely contributing, that I should not feel justified were I to omit all notice of it. Mr. C. N. Palmer had requested him to take into consideration a new project of Mr. Charles Broderip, "for the construction of a steam-vessel upon a principle of his own, to make a voyage to the West Indies," begging that Brunel would allow his name to appear as "superintending, or consulting engineer. Brunel, in his reply, says, " as my opinion is that steam cannot do for distant navigation, I cannot take any part in any scheme " connected with such a project.

Two explanations have been suggested: either at the time that the proposition was made his mind was altogether absorbed in the full anticipation that carbonic acid gas was destined to supersede steam altogether as a motive power; or, which is far more probable, owing to the difficulty of stowing, in the vessels of that day, the necessary supply of fuel—a difficulty subsequently entirely overcome by his son,— his opinion was the result of wide investigation and accurate calculation. Indeed, this view found, some years after, a full illustration in the result of the trial between the Sirius of 700 tons burden and the Great

Western of 1350 tons, in their passage across the Atlantic.

The Great Western took her departure from Bristol *eight days after* the Sirius had left Cork, and arrived at New York *only seven hours after* the Sirius; and while the former still retained sufficient coal for five days' consumption (125 tons), the latter had not only exhausted all hers, but had consumed her spare spars and furniture to boot.

In the year 1824 was revived the project of forming such a communication across the river Thames as should offer no impediment to the navigation. This problem was one which had more than once engaged the public attention, and had already occupied Brunel's mind. We have seen that in 1818 he had taken out a patent for a gigantic boring machine, which he first proposed to put in operation under the Neva at St. Petersburg.

From the universal interest which this unique structure has created, and with which it is still regarded, I prefer to break the chronological order of his labours, by first noticing less important works undertaken during the earlier period of the tunnel operations, in order that a consecutive and detailed account may afterwards be presented of the events connected with an undertaking for which there is no parallel, and with which the name of Brunel is so entirely identified.

A variety of inventions and projects were submitted to Brunel during this year (1824). Brunton's chain cable and anchor; Taylor's scheme for applying several engines yoked together horizontally, for working the Mexican mines; Ellison and Bloxam's (assignees of Fourdrinier) paper-making machine, the patent for which had been sought to be invalidated by Elsie and West, but which was successfully supported by Brunel's evidence ; Chapman's paddle-wheel ; Macdonald and Hesslop's project for opening a communication across the Isthmus of Panama. This project excited Brunel's interest, and he took much pains to obtain such necessary information as would enable him to form an opinion. The only charts which he could find were copies from old Spanish surveys, which proved of little value, whilst Humboldt's account was not sufficiently specific. From the data which he collected, Brunel could only recommend that a railroad should precede any attempt to establish the more desirable communication by canal.

The labours of Colonel Lloyd, Captain Felmarc, Mr. Wheelwright, M. Napoleon Garella, Mr. Frederick Kelley, and others, having in the course of twenty-six years supplied the data, a railway was constructed from Panama on the Pacific, to Navy Bay on the Caribbean Sea. It passes over a summit level of 254 feet, with gradients not exceeding 1 in 53 on the north, and 1 in 60 on the south.

This important work was commenced in January 1850, and was, by American enterprise, completed in five years, under the energetic direction of Mr. Tatten. In *three hours* passengers are transported from the Pacific to the Atlantic Ocean, and merchandise in *five ;* and thus the transit to England, which round Cape Horn occupied not less than *six months*, is now performed in *one ;* nor can this be considered as more than an earnest of what may yet be accomplished to meet the requirements of commerce.*

A project for the formation of a subterraneous aqueduct between Hammersmith and Hampstead occupied

* Minutes of proceedings of Institution of Civil Engineers, 1849–1856 ; Bollaert's researches in South America.

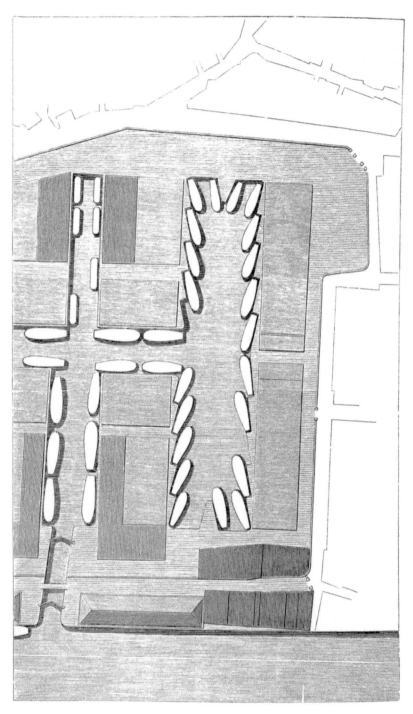

PLAN FOR WET DOCK.

much of Brunel's attention this year, as also did designs for a suspension bridge at Kingston, for which he supplied detailed estimates. The authorities having, however, ultimately determined upon erecting a more permanent structure in stone, Brunel's designs were not adopted. To the Huddersfield Canal Company he furnished designs for two suspension bridges, and to the St. Saviour's Dock Company plans for improving the works at Bermondsey, and for the construction of new docks to be called the South London. Four plans were submitted, one of them presented the novelty of arrangement here represented (see plate).

The advantages proposed to be derived from the construction were thus described : —

" Each ship has a distinct and unobstructed berth, of easy access ; and in consequence of which no ship can ever come in contact with another. The ships are more easily disengaged from the wharf, whether for the purpose of going out of the dock, or, more particularly, in case of danger from fire."

The arrangement further allows the stowage of at least ten ships in the space occupied in the other docks by seven.

The other arrangements comprehended underground warehouses, and the contiguity of the warehouses and wharf, together with cranes so situated that the goods " could be deposited in a convenient situation to be raised into the warehouses above, or lowered at once into the warehouses below." Thus " a saving of labour and time would be effected, and the service of the dock conducted cheaper, which, whether considered as a source of profit or as a means of competition, is of considerable importance."

Whether swayed by certain objections which had

been urged against the design, or intimidated by the novelty of the projection, the directors, after much deliberation, declined to adopt the scheme. During the progress of his negotiations, Brunel records in his journal some curious evidence of the jealousy which was exhibited by one of his professional contemporaries; not only for his having presumed to enter into competition for docks, when he had never executed any works of the kind in the country, but for having been successful; thereby usurping the place of those claiming a prescriptive right in works of that character.

With two great projects of the day he declined to be connected. One, the formation of a ship canal between London and Portsmouth; another, the application of locomotive steam engines to the propulsion of carriages on common roads. In the year 1825 Brunel's time seems to have become almost entirely absorbed by the operations at the Thames Tunnel, and by the carbonic acid gas machine. Still I find him in connection with M. Gay-Lussac, relative to a project for refining tallow for making candles, and in correspondence with M. Gossier, relative to the formation of railways in France, in which he strongly recommends that the result of the experiments in England should be first ascertained; but that should there be a disposition to proceed at once, then the line best calculated to afford a remunerative return, was, in his opinion, that between Paris and Rouen, because, he says, " the road is bad and the navigation worse."

In June of this year an Act of Parliament was obtained for the erection of a new bridge over the river Dee at Chester. The old bridge was no longer calculated to accommodate the increasing intercourse with Holyhead. Seven irregular and unsymmetrical arches of the fourteenth and sixteenth centuries, supported on enormous piers and abutments, were not only unsightly, but, owing to the steep and crooked approaches, the passage over the bridge was absolutely dangerous.

A bold and elegant design had been submitted to the commissioners appointed under the Act, by Mr. Harrison, the accomplished architect of the ancient city, for a structure in stone, exceeding in dimensions anything that had hitherto been accomplished. The arch was to be a segment of a circle of 140 ft. radius, the span or chord 200 ft., the rise or versed sine 42 ft., and the intrados to be 54 ft. above low water mark ; but in consequence of the great age of Mr. Harrison, then in his eighty-second year, it was impossible for him to take any active part in the execution of the work.

The cost of this noble structure was at first estimated at 30,000*l.* Mr. George Harrison, one of the commissioners, who entertained a high respect for Brunel's genius, though little acquainted with Brunel personally, having been much struck with some observations which fell from him, relative to the expense which a structure of the magnitude described would be in cut stone, and with the suggestion that it could be perfectly well executed for one third of the money in rubble masonry, communicated Brunel's views to his brother commissioners, and was authorised by them to obtain such detailed specifications and estimates as would justify them in going to the Loan Commissioners for the necessary funds. " The idea of a *stone rubble bridge* is so new in this country," says Mr. Harrison, " and the nature of cements so little understood, that you will not think it surprising our committee should wish as much information laid before them as possible, before they can adapt their ideas to a mode of building, which, for want of better information perhaps, appears to them somewhat problematical."

Mr. H. further stated that the limit of expenditure would be 31,000*l*., and that a Mr. Trubshaw had proposed to execute the work for 31,500*l*. in stone. This Brunel declared to be simply impossible, and that, in his opinion, the cost of such a design executed in solid masonry would not be less than 45,000*l*. The apprehensions of the commissioners were, however, not to be overcome, and, accordingly, their resolution of the 5th January 1826 announced, " that the consideration of the bridge upon Mr. Brunel's plan be abandoned. That the thanks of the meeting be given to Mr. Brunel for the valuable information he has communicated respecting the construction of a bridge of rubble, and that he be informed that the commissioners regret that they cannot adopt his suggestions, as, from the information they have received of the cost of erecting a bridge of solid masonry, they are of opinion that the saving of expense, by adopting his plan, would not be commensurate with the risk run in applying a totally new mode of construction to the erection of a bridge of such great dimensions."

In vain did Brunel urge the universal practice of building with rubble throughout the Roman empire. Walls, domes, cylindrical vaults and aqueducts, palaces, theatres, cisterns, all descriptions of buildings, in short, were executed in this substance.

In the East, the most splendid examples of the use of rubble are to be found. In Egypt, in China, and in India, the most gigantic works have been executed in rubble. The vault of the hall of the college at Lucknow, offers a splendid example of this kind of structure. It is 162 ft. long, and 53 ft. broad, with two verandahs 26 and 27 ft. wide, and will bear comparison with the boldest and most scientific vaulting of ancient or modern times. The evidence which French engineers have recently afforded of the value of rubble building, fully justifies the bold project of Brunel, although the adoption of the system in France, does not date farther back than 1849. Amongst the many admirable structures erected since that time, the most remarkable is the "Pont de l'Alma, by M. Darcel. This consists of three elliptical arches ; two of 126·23 ft. span, and one of 141·4 ft., with a radius of curvature of 176·3 ft., and which, as Mr. George Rennie, in his highly interesting and instructive paper, read at the Institution of Civil Engineers, liberally admits, " has accomplished in rubble and cement as much and even more than is accomplished in the granite arches of new London bridge." *

Upon the rejection of his proposition, Brunel submitted an estimate for a brick structure of the same dimensions, founded on accurate experiments which he had made in connection with the Thames Tunnel, and which would also bring the cost within the limit which the committee had prescribed ; but as the opinion of Mr. Telford, who occupied at that time the position of consulting engineer to the Loan Commissioners, was believed to be unfavourable to either rubble or brick, and as the commissioners themselves were unwilling to incur the responsibility of so novel an undertaking, Brunel's services were declined. I shall only add that this bridge, commenced in 1827, was ultimately com-

* On the Employment of Rubble, Béton, or Concrete, in Works of Engineering and Architecture, by George Rennie, M. Inst. C. E., F.R.S., May 5th, 1857.

pleted in 1832, at a cost of 49,900*l.*, including 7,500*l.* for the embanked approaches*, and is a truly noble monument to the enterprise of the good citizens of Chester, the boldness of the architect, Mr. Harrison, whose design was retained, and the skill and perseverance of the engineer and the contractor, Messrs. Hartley and Trubshaw.

Brunel was also called upon this year to examine and report upon the practicability of constructing a ship canal of about thirteen miles in length between Fowey harbour, on the south coast of Cornwall, and the Padstow river on the north. So long ago as 1794 Messrs. Bentley and Bolton had prepared surveys and estimates for a similar work, and in 1796, Mr. John Rennie submitted a project on a reduced scale to Sir William Molesworth, Bart., of Pencarrow. These movements tended to show the value attached to a direct intercommunication between the north and south coast of Cornwall.

The once celebrated harbour of Fowey seems to have undergone the vicissitudes which sometimes attend special localities. In the reign of Edward III. it was ranked as one of the first ports of the kingdom; having actually furnished forty-seven vessels to that monarch, when about to undertake the siege of Calais. In more recent times an 84-gun ship, the San Nicolas, was taken in at low water, and anchored within a few yards of the land. The harbour is esteemed the best outlet to the westward of all the ports in the west of England, and is capable of affording safe anchorage to vessels of 1000 tons burden, with three fathoms of water at the entrance at low water, deepening

---

* Transactions of the Institution of Civil Engineers, vol. i.

considerably within. Padstow is also remarkable as offering the only place of shelter on the north Cornwall coast. To unite these ports was considered to be matter of national, as well as local, importance. The tedious and sometimes dangerous navigation round the Land's End would be avoided, and the variety of produce which could be interchanged with advantage between the English and St. George's Channels, offered the prospect of a large return to the shareholders. The shelly sands of the northern coast of Cornwall, with the lime coast of Devon, would form an important manure for the old red sandstone and granite of the south and west; while the various ores of Cornwall would find easy transport to the smelting furnaces of South Wales.

The only engineering impediment was the high lands about Lanhydrock, which Brunel proposed to penetrate by a tunnel of "large and suitable dimensions;" and the cost of the whole undertaking he estimated at 450,000*l.*

Notwithstanding the advantages offered, it was, however, eventually considered that they were not sufficient to overcome the objections that vessels would only resort to the canal in bad weather, and that in good weather the same wind which would enable vessels from Swansea and Cardiff to clear Hartland Point, would also permit them to clear the Land's End; while, on the other side, vessels would not abandon the Channel to embay themselves at Fowey.

In September of this year (1825), Brunel received a kind and friendly communication from the Duchess of Somerset to visit Berry Pomeroy, for the purpose of examining the locality with the view of supplying designs for a new bridge at Totness over the Dart.

" You are always so obliging," writes Her Grace,
" that in the midst of your important business I venture to trouble you upon a concern in which I would
ask your advice, well knowing that upon the subject
in question *no* opinion is more valuable; indeed, were
you not so deeply involved *underground,* I should
propose your putting yourself into one of the coaches
and coming to us to Berry, where we should be delighted to see you."

Brunel, in reply, sent a very elegant design for a
bridge of two arches to be constructed in stone.

A good deal of Brunel's attention in the autumn of
this year was occupied by an apparently trivial matter.
An exploring expedition had been undertaken to the
Bay of Vigo, where, in 1702, a number of valuable
Spanish galleons, under a French convoy commanded
by Chateau Renault, had taken refuge from the combined English and Dutch fleet; but where, being unable
to resist an attack, the Spaniards, after having secured
the best part of their plate and merchandise, resolved
to burn and sink their vessels.   Before, however, they
could accomplish their object, ten ships of war and
eleven galleons were captured.   Still, considerable treasure was supposed to have been sunk, and it was with
a view to recover that treasure that the expedition was
now undertaken.   One of the most important means
to be employed was a CRANE destined to work the
diving-bell, and which had already been fitted up in
the brig Enterprise; but doubts having arisen as to its
safety, Brunel's opinion was taken.   The number of
alterations which this simple machine had to undergo,
before Brunel could pronounce it competent to fulfil
the duties which it was destined to perform, amply
justified the appeal to his judgment.   I may add, that

the Vigo Bay expedition appears to have been one of the many projects of this prolific year, by which unprincipled schemers sought to excite the cupidity of those who would " make haste to be rich."

That *auri sacra fames* spread itself like a plague through the land, embracing all classes in its fatal grasp ; and though those in charge of the exploring vessel to Vigo, did make some efforts to fulfil the objects of the expedition, the scheme itself proved a total failure. Mr. Abbinett, writing to Mr. James Forrest (Secretary to the Institution of Civil Engineers) in August last, states, that being well acquainted with the agent and the steward of the Enterprise brig, he had an opportunity of knowing the result of the expedition. All he saw was some pieces of timber and some dinner plates, though the captain of the vessel stated that they " worked to the bottom, and all through one of the ships, and found nothing." This want of success did not deter others from making further attempts, which continued from time to time up to 1845, with no better result.

On the 21st of August, 1826, Brunel received an invitation from Liverpool to design " a landing for passengers from steamboats, so as to be equally convenient at high and low water ; " and on the 26th Mr. Foster writes, that it is the request of the Mayor, Mr. Bourn, and of the Committee of the Common Council, that " you come down and inspect personally the site of the proposed landing-places, as it would be almost impossible to explain, by any plans or description, the inconvenience the public at present labour under from the number of steamboats which frequent this port, and the frequency of their arrival, as also the immense number of passengers which each boat brings." On the 1st of September, Brunel reached Liverpool, travelling the last fifty-four miles, he says, in seven hours, with one hour's stoppage; therefore, at the rate of nine miles an hour. The following morning he witnessed a scene of confusion, involving an amount of personal inconvenience if not danger, at the principal landing-place, which he finds it impossible to describe; but which had the effect of stimulating his inventive faculty to supply at once a want so imperatively demanded. By November, the drawings for a floating

ORIGINAL DESIGN FOR FLOATING PIER AND APPROACH.

pier were completed, and on the 1st of December they were submitted by his son, together with a working model, to the Committee of the Common Council. Notwithstanding our familiarity with this valuable appliance, the accompanying sketch from Brunel's original drawing may prove of interest to those who desire to mark the steps by which our social and commercial intercourse has been facilitated.

The principle here established by Brunel gradually came to be applied generally throughout the kingdom. At Liverpool, where the population had increased from 118,972 in 1821, to 252,236 in 1851, the demand for farther accommodation for landing became imperative, and accordingly in 1847, a new landing stage was designed by Sir W. Cubitt, 507 feet long and 80 feet broad, connected with St. George's Pier by two iron bridges 150 feet in length and 17 feet in width, at a cost of 40,000*l.*; but this proving still insufficient, another was constructed by the same distinguished engineer, of still greater magnitude. Sixty-three pontoons, supporting a platform which extends upwards of 1000 feet along the Prince's Dock, and connected with the esplanade by four bridges, at a distance of 100 feet from the wall, may now be seen; while horizontal cables of iron, two inches in diameter, retain the vast structure in position independently of tides, which vary in elevation from 20 to 35 feet.

CHAPTER XIV.

1824–1825.

THAMES TUNNEL — EARLY ATTEMPTS — DODD, 1798 — VAZIE, 1802
— TREVITHICK, 1807 — HAWKINS — TEREDO NAVALIS — ORIGIN OF
THE SHIELD — FORMATION OF THE THAMES TUNNEL COMPANY,
1824 — APPOINTED ENGINEER — CONSTRUCTION OF THE SHAFT,
1825 — DESCRIPTION OF SHIELD — NATURE OF THE BED OF THE
RIVER.

IT is with much hesitation, that I now approach the great and final effort of Brunel's ingenuity and skill. Fully aware that impressions made upon the mind when under the influence of enthusiastic excitement or anxious responsibility, are apt to be over estimated, I might well doubt my ability to render a faithful account of events, in which I have been personally so much interested. Fortunately, however, I am enabled to correct such impressions by reference to records of almost unexampled detail. Although the novelty of the undertaking has now passed away, a large amount of interest must still attach to a work which bears so strongly the impress of a high mechanical intelligence, supported by an amount of courage, energy and perseverance rarely demanded of the civil engineer.

We have seen that Brunel's mind had been long directed to a system of tunnelling in alluvial ground which should supply the miner with a continuous loco-

motive protection, and that in 1818, he took out a
patent for a machine with this object.

Before I enter upon a record of Brunel's labours, in
this new department of engineering, it may be interesting to cast a retrospective glance at the nature of the
precedents which had been established, and the character of the works which had been projected.

The honour of having first conceived the bold project of uniting the counties of Kent and Essex by a
subaqueous communication, is generally conceded to
Mr. Dodd, an engineer of considerable eminence.

In the preface to his Report, addressed to the nobility and gentry of Essex and Kent, May 1798, he
says :—

" From the importance of a communication between
the towns of North and South Shields, which was under
my constant view, and where no bridge could possibly
be constructed, my mind happily thought upon the
scheme of making a subterraneous, and I may say
a subaqueous passage to accomplish this desirable
purpose."

" In the course of my professional travelling," he
adds, " I observed the want of a grand uninterrupted
line of communication in the south-east part of the
kingdom, which would easily be obtained if the river
Thames could be conveniently passed."

The points selected by Mr. Dodd to cross the river
were Gravesend and Tilbury ; the distance, 900 yards ;
the tunnel to be cylindrical and 16 feet diameter ; the
cost, 15,995_l._

In 1802, Mr. Vazie selected the part of the river
from Rotherhithe to Limehouse. His project met with
supporters, and a company was formed. To explore
the ground it was considered absolutely necessary that

a drift-way should be executed, which might subsequently form the drain to the greater work of the tunnel. A shaft was accordingly sunk, *eleven feet* in diameter, to a depth of *forty-two feet.* This operation occupied a whole season in consequence of the false economy of the directors, who were tempted to substitute a 14-horse power steam-engine of imperfect construction for a 50-horse required by the engineer.

This brought its own punishment, for the water from a stratum of gravel overpowering the engine brought the works to a stand-still; 7000*l.* had been expended, and all would have been lost, had not Vazie prevailed upon a principal proprietor to undertake to sink the shaft to its required depth. By means of a caisson the bed of gravel was passed through, and Vazie, having sunk the shaft seventy-six feet below high-water, was prepared to proceed with the drift-way. But again the directors hesitated, and the works were suspended, "until the opinion of a professional man of eminence be taken on the various matters respecting it." Mr. Rennie and Mr. Chapman were consulted, but as their opinions did not coincide, Mr. Trevithick was invited to undertake the work, and a contract was entered into with him, "for superintending and directing the execution of the drift-way." On the 17th August, 1807, this important work was commenced; Trevithick as the engineer, Vazie as the executive. The dimensions of this drift-way were: height, 5 feet; width at bottom, 3 feet; at top, 2 feet 6 inches. In the beginning of October, when 394 feet had been accomplished, difficulties arose and the works were interrupted. Misunderstanding having arisen between Vazie and Trevithick, Vazie was removed from his responsible situation, not without a protest on his part.

From the statement addressed by him to the proprietors, we learn, that considerable departures from his original design had been sanctioned by the directors, which he predicated would lead to ultimate failure. He complained of the manner in which he had been treated after a devotion of four years to the interests of the company, during which time, he says, " I never slept one night from the works ;" and adds, that it had been " by no means unusual for me to remain *forty hours* without rest."

The execution of the work now devolved altogether on Trevithick, who was to receive 1000*l*. premium if he succeeded. It continued to progress favourably to the 22nd of December ; by that time 953 feet had been completed, when a quicksand was encountered ; but the drift having ascended above a stratum sufficiently dense to be designated rock, the roof broke down, although there was not less than thirty feet of ground between the top of the drift and the river.

This difficulty was, however, courageously overcome, and, by the 26th January, 1028 feet had been accomplished. The bed of the river now gave way, and the water inundated the drift. Clay in bags was thrown into the hole, and the interstices filled with gravel and earth. The hole was stanched, the work was resumed, and 70 feet more were added to the excavation, when the roof broke down a second time, and sand and water entered the drift way with such violence that in about a quarter of an hour the water rose in the shaft nearly to the top. Again the bed of the river was made good with clay and gravel, the water pumped out, and the works proceeded ; pipes being driven in various directions into the ground for the purpose of allowing the water to pass free from sand. Under great danger and difficulty twenty feet more were accomplished; but the bursts of water and ground became so unmanageable, that the face of the drift was ultimately timbered up, and the work abandoned.

It is no doubt true that many excavations have been made under deep rivers, nay, under the sea itself, in the progress of mining; but we have here the fact that under the bed of the Thames, practical miners, selected for their skill and experience, uniting in a high degree physical and mental energy and power, and confident in their own resources, had not been able, with all their efforts in the course of five years, to complete their *drain;* not a single brick had been laid, and everything which had been done was irretrievably lost. The opinions of scientific and practical men were now sought for, and a premium was offered of 500*l.* for a plan that could be acted upon. A vast number were received, of which forty-nine were selected, and in 1809 submitted to Dr. Hutton, the celebrated mathematician, and to Mr. Jessop, one of the most accomplished civil engineers of the day, for an opinion and report. The conclusion to which these gentlemen came is thus stated: "Though we cannot presume to set limits to the ingenuity of other men, we must confess that, under the circumstances which have been so clearly represented to us, we consider that an underground tunnel, which would be useful to the public and beneficial to the adventurers, is *impracticable.*"

There is little doubt but that the conclusion at which those gentlemen had arrived, and the opinion which they expressed, had the effect of deterring the public from offering support both to schemers and enthusiasts; and so the question was set at rest, until Mr. R. F. Hawkins, in 1816, proposed a method in which he had so much confidence, and of the success of which he was so well satisfied, that he secured it to himself by patent.

Two shafts of brick in cement were to be sunk in the river two hundred feet from each shore to a sufficient depth. From these shafts the excavations were to proceed in both directions; the bottom of the shafts to form wells, from whence the water was to be removed by pumps fixed on the top of the shafts; the roof to be of cast iron. I am not aware that any attempt was ever made to realise this project. It appears, however, to have had the effect of rousing the attention of Brunel to the question. At the time when Mr. Hawkins' project was put forward, Brunel was completing his works at Chatham, and one day, as he himself related to me, when passing through the dockyard, his attention was attracted to an old piece of ship timber which had been perforated by that well known destroyer of timber—the *Teredo navalis.* He examined the perforations, and subsequently the animal. He found it armed with a pair of strong shelly valves which enveloped its anterior integuments, and that, with its foot as a fulcrum, a rotatory motion was given by powerful muscles to the valves, which, acting on the wood like an auger, penetrated gradually but surely, and that as the particles were removed, they were passed through a longitudinal fissure in the foot, which formed a canal to the mouth, and so were engorged. To imitate the action of this animal became Brunel's study. " From these ideas," said he, " I propose to proceed by slow and certain methods, which, when compared with the progress of works of art, will be found to be much more expeditious in the end." As we have seen, he so far matured his ideas as to take out a patent in 1818 for a machine of iron forming auger-like cells for the miners ; and which should be forced forward with a rotatory motion by hydraulic presses, displacing only so much ground as the machine would occupy in its place. For a large excavation many of the cells were to be connected together, but at the same time provided with an independent action. Farther consideration showed that a circular structure was not adapted to the free exercise of the power of man, and the angular form was substituted. Brunel, urged by one of the active promoters of the archway enterprise (Mr. I. W. Tate), now devoted himself to the development of his ideas. At the close of 1823 he communicated his views to a few friends, to whom he fully explained his plans. Men of experience, acquainted with difficulties which had overpowered the former efforts of Vazie and Trevithick, were amongst the earliest advocates of Brunel's plans.

On the 18th February, 1824, the first general meeting of those interested in the project of forming a tunnel from Rotherhithe to Wapping was held at the City of London Tavern, W. Smith, Esq., M.P., in the chair.

The means by which such a work could be accomplished were explained. The received notion that a preliminary drift-way was indispensable, was found to be abandoned, and an excavation sufficient to receive a double archway of full dimensions was boldly proposed. The accompanying plate represents the comparative area of the two excavations ; the smaller, that of the drift-way of Trevithick ; the larger, that of the Thames Tunnel of Brunel.

That an area of 630 feet (afterwards increased to 850 feet) should be undertaken by one who had no experience in mining operations, and in the very locality which had already defeated the best efforts of skilled miners, was sufficient to excite the astonishment of the engineering world ; but as the mode of effecting this gigantic excavation began to be understood, and the unrivalled mechanical capability of the projector was remembered, a feeling of confidence spread through the scientific and practical mind of the country, which resulted in the formation of a company. The number of shares authorised to be issued was 4000, which at 50l. a share would have amounted to 200,000l. But of these only 3874 were subscribed for, and ultimately, in consequence of 276 defaulters, the total sum actually subscribed was 179,900l.

On the 18th February, 1824, the first general meeting was held at the City of London Tavern, and 2128 shares of 50l. each were at once subscribed for. The name of George Hyde Wollaston heading the list for 500 shares, followed by that of his distinguished brother, Dr. Wollaston. A committee was formed, which included the names of men known for their mechanical acquirements and professional position ; Sir Edward Codrington, Messrs. Bramah, Donkin, Gray, Martin, Ritchie, B. Shaw, W. Smith, M.P., W. Taylor, and G. H. Wollaston.

Mr. Montague, the city surveyor, was appointed to survey and estimate the value of property required, and to Messrs. Joliffe and Banks was intrusted the responsibility of making the necessary soundings and borings in the river, which were to guide Brunel in determining the special character of his proposed machine.

On the 25th June, 1824, the Bill for the incorporation of the company having passed without opposition through both houses of parliament, received the royal assent, and on the 20th July a general meet-

P

ing of the shareholders took place at the City of London Tavern, W. Smith, Esq. M.P., in the chair.

In the report submitted by the committee, it was announced " that they have now the satisfaction to inform you that the result of *thirty-nine borings* made upon two parallel lines across the river has fully confirmed the expectations previously formed, there having been found upon each a *stratum of strong blue clay of sufficient depth to ensure the safety of the intended tunnel.*"

" The ground on the Surrey side of the river, near to Rotherhithe Church, was also bored, and a deep well being sunk on the north side, for a parochial purpose, gave a result of the most encouraging nature." The Report goes on to state that " in compliance with instructions from the subscribers at the original meeting, your directors have made arrangements with Mr. Brunel for the use of his patent, for which they have agreed to pay him 5000*l.* when the body of the tunnel shall be securely effected and carried sixty feet beyond each embankment of the river, and a further and final sum of 5000*l.* when the first public toll under the act of parliament shall have been received for the use of the proprietors."

" To effect the work Mr. Brunel has received the appointment of engineer to the company, with the salary of 1000*l.* per annum for a period of three years, the utmost limit which the directors contemplate as necessary for the execution of the work ; the whole of which sum the directors have agreed to give him in case the work should be accomplished to their satisfaction at an earlier period."

These arrangements permitted Brunel to select a residence more in accordance with his improved means ; he, therefore, proceeded to transfer his office from the

Poultry, and his family from Battersea, to No. 30 Bridge Street, Blackfriars. Nearly the whole of his time was now occupied in perfecting his designs, and in preparing the working drawings for a shield, which should be specially applicable to the ground that had been reported by the surveyor as forming the bed of the river; in determining the best means of sinking the shaft, which was to form the prelude to his subfluvial operations, and in making accurate experiments on the adhesive properties of Roman cement.

On the 16th of February, 1825, arrangements were so far advanced as to permit the ground to be cleared for the construction of a shaft *fifty feet* in diameter, and *forty-two feet* in height.

The locality selected, was contiguous to Cow Court, Rotherhithe, and at a distance of one hundred and forty-one feet from the river wharf.

A circular curb of timber was laid on the flange of a curb of cast-iron, two feet six inches deep, formed in sections; the inner rim being supported by folding wedges laid on the notched heads of short piles fourteen inches square.

The piles were driven into the ground by a monkey of 296 lbs. weight with a fall of six feet until they were found to offer a uniform resistance.

Things being thus prepared, the first stone was formally laid by the chairman of the company, W. Smith, Esq., M.P., on the 2nd of March.

Upwards of two hundred persons partook of a sumptuous collation; the bells from the steeple of Rotherhithe rang out their joyful acclamations, and success to the undertaking was echoed from a thousand voices. On the following day the brickwork of the shaft was commenced. On the curb a wall 7 feet 2 inches high was erected, composed of dry bricks, for the purpose of more uniformly pressing down the curb and securing an equality in the foundation for the tower or shaft. The weight thus obtained was 190 tons, and this, with the timber and iron curbs and bolts, amounted to 215 tons. So uniform was the bearing that in twenty-four hours the settlement did not exceed the sixteenth part of an inch. On the 18th of May the permanent brickwork in cement was commenced. The dry bricks being removed, two walls, one brick thick, were built with Roman cement, the inner faces being coated with cement composed of Roman cement three parts, and sand two. The trough thus formed was filled with bricks well grouted with cement.

From the timber curb ascended 48 wrought iron bolts one inch diameter, which were built into the brickwork, and which at the top were united to another timber curb by nuts and screws, thus binding the whole work into one solid mass. Iron hoops, and timber bonds nine inches deep and half an inch thick, were also let into the brickwork at intervals as the work was brought up.

In three weeks the shaft or tower, weighing 910 tons, was completed, each bricklayer laying one thousand bricks a day. During that time it had descended uniformly half an inch. The next step was to ease and gradually to strike the wedges under the curb. This was accomplished on the 15th and 16th of April, and the curb was brought to bear on the piles. The shaft was now found to have settled $1\frac{3}{8}$ inches on one side and $1\frac{1}{8}$ on the other; the perpendicularity was speedily restored by driving three of the piles at the higher side. Three days afterwards, the whole of the piles were driven down two by two opposite to one another to

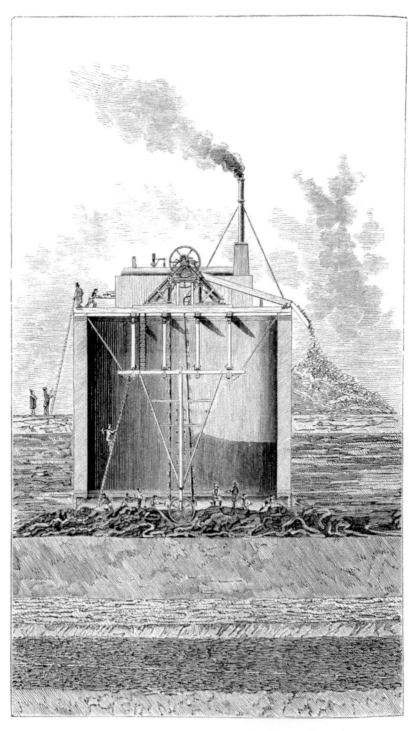

SECTION OF SHAFT IN PROCESS OF BEING SUNK

allow the curb to bear entirely on the gravel. The piles were then removed, when the whole was found to have sunk uniformly two inches and a quarter. A well was sunk contiguous to the shaft, not only to act as a drain, but also the more readily to determine the nature of the ground, and to anticipate any difficulties on that score which might arise in sinking the shaft.

Workmen were now placed inside the shaft, and by loosening the gravel all round, and casting it towards the centre, it was thence raised by a windlass and buckets to the surface, and the shaft descended by its own weight.

On the 21st of April it sank $8\frac{3}{4}$ inches, quite vertically. On the 22nd His Grace the Duke of Wellington, accompanied by Lord Somerset and General Ponsonby, honoured the works with a visit, descended into the shaft, and witnessed with intense interest the operations.

The crunching sound produced by the entrance of the iron curb into the gravel in a circumference of 157 feet being reverberated from the walls of the tower, had a striking, not to say startling, effect; while it tended to exalt the impression which the magnitude of the operation was so well calculated to inspire. Occasionally the descent was checked by the binding of the gravel round the curb ; but water being introduced to loosen the gravel, the interruption from that cause was quickly overcome. In the progress also a part of the iron curb was broken ; but in consequence of the facility which the construction offered of removing any portion without injury to the contiguous parts, there was no difficulty in effecting repairs. The strength and cohesion of the structure was, however, soon to be severely tested. The stratum of gravel was found not to be uniform. On the 29th April a change was observed. Suddenly the tower descended *seven inches* on the east side where it had entered some soft ground, while the descent on the west, where the ground still remained hard with large stones embedded, was only *three and a half inches*. The surge was alarming, but so admirably was the structure bound together that no injury was sustained. Considerable delay now occurred in consequence of the rapidity with which the preliminary operations had been conducted under the immediate superintendence of Mr. Armstrong the resident engineer, aided most effectually by young Isambard Brunel, then only in his nineteenth year ; but who already exhibited an amount of physical energy and intellectual vigour which gave no ordinary promise of future greatness. The steam-engine was not forthcoming which was to work the pumps, as well as the buckets, by which the excavated soil was removed, and which in consequence of a considerable influx of water, became of economic importance. In effect upwards of one hundred men formed but an imperfect substitute for a fourteen horse-power steam-engine. At length the operations had to be suspended until the steam engine could be applied. Some days were thus lost during which time the water rose fifteen feet in the shaft. By the 16th May a steam-engine was fixed on the top of the shaft, and the work proceeded with renewed vigour. During the progress of sinking, which varied with the character of the ground from three inches to twenty inches in the day, it was often found advantageous to permit the water to rise some feet in the shaft whereby the soil round the edge of the curb became loosened in a more equal manner than could be accomplished by the pick of the miner.

By the 3rd June the shaft having been sunk to within two feet of the required depth, and although the ground had been cleared from beneath the curb for that distance, the pressure of the surrounding ground still held it fast. By loading the top first with 8,000 bricks and subsequently with 50,000, and by permitting the water to rise, the object was finally accomplished on the 6th of June. In compliance with the Act of Parliament the top of the shaft was to be three feet above Trinity high water mark, and this was effected within seven inches.

Brunel now proceeded to underpin the shaft and to remove the curbs, a work of no little difficulty where the ground in some parts seemed so loose as to be always ready to run in. On the 7th of June the operation was commenced, and was to proceed night and day. Unfortunately, on the 8th the steam-engine, never good, and only employed because the engine specially ordered was not ready, demanded immediate repairs. This compelled the suspension of all operations, and permitted the water to rise in the shaft twenty-one feet. As a consequence, the silty ground gave way on the north side and the gravel ran in, leaving a considerable cavity behind the shaft. Sheet piles were now driven horizontally under the curb which proved effective in checking the gravel, and a quantity of shavings being pressed into the cavity, the water was disengaged from the gravel and the work proceeded the following day. On the 14th and 23rd a similar cause produced similar results, and again the same means were resorted to for overcoming the difficulty.

By the 7th July, the underpinning had been successfully accomplished to within two feet of the lowest point, viz. 62 feet 6 inches below high water mark, Trinity datum, leaving an opening of 36 feet in width to receive the shield, which was secured by timbering. To complete those two last feet rubble stone work was employed, well grouted with mortar, composed of 1 part Roman cement, 2 lime (lias), 3 sand, and which was also used in the construction of the inverted arch at the bottom of the shaft.

It was a subject of regret to Brunel that he had not provided for sinking the shaft the total depth required; not only as being a less expensive, but more expeditious and far safer mode of operation than underpinning.

During these novel and interesting operations the works were visited by crowds of persons, sometimes to the great inconvenience of the workmen. Amongst those belonging to the higher ranks I find the names of their Royal Highnesses the Duke and Duchess of Cambridge; His Royal Highness the Duke of Gloucester; Prince Leopold; the Duke of Northumberland; Mr. Peel (Sir Robert) and his lady, &c.

So admirable had been the arrangements, and so continuous and successful the progress of these gigantic operations, that the men employed began to treat them as ordinary work, and to give way to indulgences only too common amongst their class, but from which they had hitherto been restrained by the very novelty of their engagements. The confidence, however, which had been gained proved unfortunately fatal to one of the gangers named Painter, who, coming intoxicated to the works on the night of July 12th, fell from the top of the shaft to the bottom: strange to say, he survived the concussion twelve hours.

While the works of the shaft were being proceeded with, the contiguous well in the yard was sunk below the bottom level of the shaft. This had the intended

effect of acting as a drain, the water naturally gravi-
tating to the well, from whence it was readily pumped
out.

The next step was the formation of a reservoir to
receive the permanent pumps. Before this could be
ventured upon, it became absolutely necessary that the
new steam-engine and pumps should be in place, the
former supplied by Maudslay, the latter by Messrs.
Taylor and Martineau. It having been observed, that
the ordinary leather-valved pump was constantly requiring repair from the facility with which the shelly
matter in the silty ground cut through the valves,
Brunel determined upon employing plunger force
pumps. These were adopted, and used during the
whole period of the tunnel operations, with perfect
success. The steam-engine was a double-acting high
pressure of 24-horse power, the cylinders inclining
at an angle of 45°.

By the 19th of August the steam engine and pumps
being in place, and successfully tested, borings were commenced in the centre of the shaft, which was 55 feet
6 inches below high water. At a distance of 14 feet or 69
feet 6 inches below high water, a stratum of sand largely
impregnated with water was met with. When opened,
the sand seemed to boil up in the pipe through which
the boring was conducted. This stratum being only
one foot in depth was passed through, nor until a depth
of 85 feet had been attained was any more water encountered. It was, therefore, determined to sink the
reservoir below the first spring, but not to touch the
second. To this end a square well with sheet piling
was commenced at the bottom of the shaft. When it
had been carried down about thirteen feet, the water
burst up with considerable force. Rubble stones were

thrown in to prevent the sand rising with the water, and as the water from two other openings into which pipes had been introduced immediately ceased to flow, and as the absolute increase of water was very small, the work proceeded without any apprehension being excited, and the piles were driven through the stratum of sand. Brunel having by this excavation for the well satisfied himself as to the exact nature of the ground on which he proposed to form the great reservoir, he proceeded to clear the ground for the curb upon which it was to be built. When it had been sunk about five feet, the ground on one side suddenly gave way, the sides of the hole remaining perpendicular. It was evident that the sand had been gradually washed away and taken up by the pumps, and it required about forty cubic feet of gravel to fill the cavity thus created. To render the operation of forming the reservoir perfectly secure, and to which Brunel attached so much importance, he directed that sheet piling should be driven round the curb, bound with sheet-iron waling, and that a second curb should be applied at the top to confine the piles in their places. As rubble stones had to be employed to check the movement of the sand, and as they had to be frequently displaced to allow the piles to be driven, this part of the operation became very tedious. When a depth of about twenty feet had been gained, or about seventy-five feet from high water mark, and as the invert was about to be commenced, another slip of ground took place behind the sheet piling, extending nearly one fourth round the opening. Into this, clay was introduced and well rammed, and being covered with rubble to prevent any washing, the work proceeded.

It had been part of Brunel's original design to have

driven a heading from the reservoir directly under the tunnel for the purpose of securing effective drainage for the great excavation. This intention he was unfortunately induced to abandon, in consequence of the strong objections raised by the directors on the score of expense, and to substitute a horizontal cast-iron pipe in its place, and a perpendicular pipe in the centre of the reservoir. As the stone-work was brought up, circular wall bonds nine inches wide and half an inch thick were laid in and grouted, similar to what had been done in the construction of the shaft. On the 11th of October the reservoir was completed in a most substantial manner, and the rubble stone-work found to have perfectly united to the gravelly bottom. A dome now closed the whole, openings being left for four pumps and for examination.

The great shield had been, during the operations above described, nearly completed by Maudslay, and on the 15th of October, two of the twelve frames of which it was composed, were lowered into the shaft and placed in position.

The accompanying isometrical sketch of one of the frames, will convey some idea of its constituent parts.

On the right is the representation of a portion of the ground in front. Against the ground the *poling boards* supported in place by the *poling screws*, resting or abutting against the *cast iron frames*. On the top the *top staves*, at the bottom the *shoes* attached to *legs*. The large propelling screws are omitted, but may be seen in the small section at page 223, which includes the stage from which the building materials employed in the upper portion of the work were supplied.

The frame will be seen to have been divided into three compartments or cells, each division being 3 feet

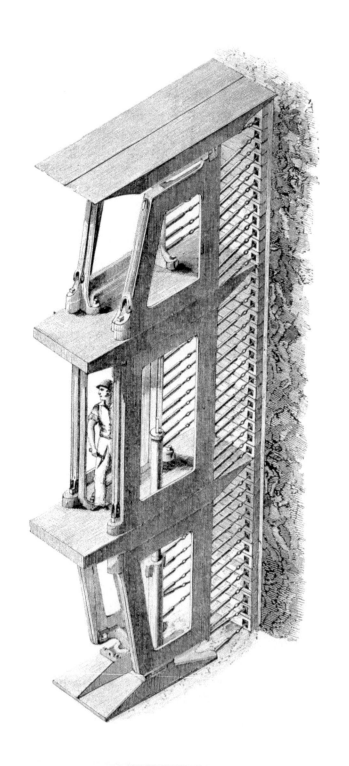

broad and 21 feet 4 inches high ; the 12 frames forming therefore together 36 cells, in which the miners worked independently of one another. Each frame stood upon two iron legs fitted with ball joints to iron shoes, and which, as will be subsequently shown, very much resembled in their action the human leg and foot. The floors of each division formed so many stages for the bricklayers, who by working with their backs to the miners, allowed the double operation of mining and bricklaying to be carried forward simultaneously. Boards, technically termed polings, placed horizontally, 3 inches thick, 6 inches wide, and in length corresponding to the width of each frame, supported the ground in front, being retained in their position by small jack screws, termed poling screws, which at the top entered shallow counter-sunk hollow plates fixed on the poling boards, and, at the base or butt-end, into notches in the front rail of the frames, and were therefore capable of resisting any pressure that might come against them. Each front rail was supplied with a double set of these notches, the reason for which will be understood when the movements of the machine are described. The support afforded to the ground above, and at the sides formed a peculiarly interesting and important portion of this machine, and was that which presented the greatest difficulty, and caused to Brunel the greatest anxiety in its projection. As before observed, the shield was divided into twelve independent parts, each of those parts or frames was again subdivided into top, middle, and bottom ; the top carried the head, this was further subdivided into three parts called staves, or slippers, or sliders, of sufficient length to cover the whole depth of the excavation. Each stave was moveable to a certain extent upon a centre or pivot, and was capable of being slid forward individually as occasion might require. The only portion of ground, therefore, which should remain at any time unprotected, was that between a stave when moved forward into the ground in front and the brickwork behind, and that only for the short time required to add the next course of bricks.

Tails were subsequently added, so that no ground should be left unprotected as the brickwork was brought close up to the tail.

The advantage derived from this arrangement of the top stave was, that as the frames could not at all times be moved forward vertically, and as the pressure of the ground was often very variable, they were thus enabled to accommodate themselves to that pressure, however unequal ; it also assisted in giving direction to the movement of the frame.

The side staves were, with the exception of the pivot, similar to the top staves. Finally, to provide against the lateral pressure, friction rollers were placed between the frames.

The modus operandi was as follows ; when a frame was to be moved, the ends of the screws abutting against the front rails of the frame, and which supported the poling boards, were transferred to the second set of notches in the rails of the neighbouring frames ; and thus, while the screws retained the polings against the ground as firmly as before, the frame was left free to be moved forward. When this operation had been performed by the miner in each cell, one of the legs, a large vertical screw, was unscrewed a few inches, which raised the shoe to which it was attached. A lever being applied to the leg, it was moved forward the required distance. The leg was then screwed down, and the other being operated upon in a similar manner, the body of the frame was in a condition to be forced forward. This movement was accomplished by means of large *horizontal screws*, which abutted against the brickwork of the arch at the top, and against the inverted arch at the bottom.

A moveable stage immediately behind the shield to receive the building materials for the upper portion of the work constituted the whole of the mechanical arrangements.

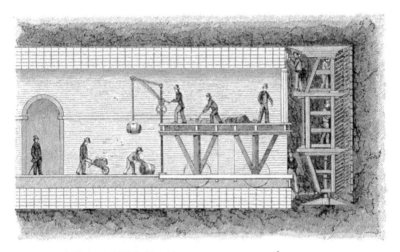

SECTION SHOWING MOVEABLE STAGE AND PROPELLING SCREWS.

A little consideration will show that the alternate frames could alone move at one time ; the intermediate frames being required to support the ground in front. Thus Nos. 1, 3, 5, 7, 9 and 11 would proceed together; while Nos. 2, 4, 6, 8, 10 and 12 gave the necessary support to the poling boards. As soon as a frame had been put forward, one poling board at a time was removed by the miner in each cell by unscrewing the poling screws. The ground in front was then excavated to the extent of nine inches ; after which, the board was again replaced, and the screws which supported it were returned to their original position in their own frames, screwed tight, and the next poling removed until the whole was worked down. The neighbouring frames were then prepared in a similar manner to advance, and the brickwork followed as before stated.

It is worthy of observation that each part of this stupendous machine was capable of being removed and repaired without in the slightest degree endangering the stability of other contiguous parts ; and though apparently complex in its structure, was still capable of being soon comprehended by the ordinary miners and labourers employed.

The character of the soil having varied very little during the first 500 feet of the excavation, may be here described, commencing at the top.

Two feet of very strong blue clay, decreasing as the dip of the tunnel exceeded that of the stratum, and which at a distance of about 250 feet no longer appeared.

Six feet of pure blue silt for 300 feet.

Six feet of blue silt with a great abundance of small bivalve shells very minutely broken, a few only perfect. These shells were sometimes found in a thick layer, leaving the silt pure. Sometimes a number of strata— shells and silt alternating—presented themselves.

A layer of about an inch thick of indurated sand, has been constantly found in the bed of silt.

Two to three feet of stone, a sort of bastard gypsum, not at any time continuous, but coming out in large lumps.

And, lastly, three to nine feet of gravel intermixed with green silt, which formed the foundation.

Here then was a totally different character of ground from that which had been represented in the report of the surveyor, and announced by the directors on the 20th of July, 1824, and in place of " a stratum of strong blue clay of sufficient depth to ensure the safety of the intended tunnel," a variety of strata was found varying in density one from another, and in themselves under different circumstances ; and it must be obvious that a machine, which was designed to operate in homogeneous blue clay, was scarcely calculated to contend against a friable sand, or sand so impregnated with water as to have become absolutely fluid.

CHAPTER XV.

1825–1827.

SHIELD IN PLACE, 1825 — PIECE WORK — ISAMBARD BRUNEL — EXAMPLE OF DIFFICULTIES — HOSTILITY OF THE CHAIRMAN — MR. GRAVATT AND MR. RILEY APPOINTED ASSISTANTS — PANICS — STRIKE — DIFFICULTIES INCREASE — DEATH OF MR. RILEY — FIRST IRRUPTION OF THE RIVER, 18TH MAY, 1827.

ON the 28th of October the twelve frames were in place, but the fitting and adjusting of the parts, and the construction of the entrance to the tunnel, occupied some weeks; meanwhile a heading or drift was opened behind the bottom of the shaft, for the purpose of removing the cast-iron curb, that the under-pinning might be completed, and enabling clay to be introduced, so that the junction of the underpinning with the shaft might be effectually protected. This involved, however, the sinking of another well from the surface, at a point where the ground was saturated with water.

On the 28th of November the shield commenced its eventful march. Two sets of miners, thirty-six in each, with the necessary complement of bricklayers, worked in what are called shifts or periods of eight hours each. The excitement to which the pro-jection and direction of these various operations had subjected Brunel, produced, on the 22nd of November, 1825, an illness so serious as to call for very active treatment, and the application of many leeches to his head, and from which he slowly recovered. The works however proceeded, and, a stratum of gravel having been encountered, water in great quantity broke into the works; considerable difficulty was also experienced in forming abutments for the large propelling screws. These could only be obtained from balks of timber thrown across the shaft; notwithstanding, by limiting the excavation to $4\frac{1}{2}$ inches, or half a brick at a time, the work progressed successfully, though slowly. To relieve the ground round the shaft from water, holes were made in the brickwork.

By the 23rd December, 1825, the first section of the double archway was completed, and the shield had entered into undisturbed ground free from water.

Scarcely, however, had 7 feet been accomplished, when the strata began to exhibit considerable inequality; as a consequence, the frame in the west corner gradually sunk below the friction rollers attached to the middle floor; but the provision made for the movement of the top staves soon restored it to position. New delays now arose from the force necessary to be applied to the large abutting screws, the threads of many of them having been stripped off.

On the 26th of January, 1826, when 14 feet of brickwork had been completed, the water burst in about the centre of the excavation with considerable force, and before the stratum could be passed through from whence it proceeded, the feed pump of the steam engine became deranged; as a consequence the engine was stopped, and the works suspended, when the water rose 12 feet in the shaft. Upon resuming the excavation (February 3rd), so much interruption was experienced from the influx of water, and so much

hesitation was exhibited by the miners, that it was considered advisable to sink a well immediately over those
frames upon which the water seemed concentrated,
when, a sort of pot-hole of gravel having been discovered with sound ground beyond, direction was given
to drive the frames through.

On the 15th of February water and gravel broke
again into the works; but, as the men had now learnt
to place more confidence in the protection afforded by
the shield, no further hesitation was exhibited, and by
introducing a leaden pipe behind the frames, the water
was drawn off to the relief of the face of the work.
By the 25th of the month the faces were free from
water, and on the 11th of March the shield had, without injury, entirely passed through ground which no
known system of tunnelling could have successfully
penetrated. Still, notwithstanding the proof which
had been thus afforded of the value of the shield, an
unfavourable view of the whole project had taken possession of the mind of the chairman of the company.
The shield was pronounced unnecessary—a delusion in
short, the construction of the brickwork false in
principle; and although in these opinions few con-
curred, still such hostility distressed Brunel, distracted
his mind, and had nearly produced another serious
illness. The successful application of his shield, however, in resisting casualties for which it had never been
designed, soon restored his natural cheerfulness, and
enabled him to bear with more equanimity the animad-
versions to which he was exposed. Unfortunately for
Brunel, the resident engineer was not now found to be
equal to the novel duties demanded from him, nor
could all the energy and intelligence of his valuable
assistant, Isambard Brunel, supply the deficiency.

Early in April, Mr. Armstrong broke down, and Brunel himself was confined to his bed for a week. The whole direction of the undertaking was then cast upon a young man scarcely twenty years of age, by whom nights as well as days were passed at the works. In two months the rate of progress had been about 8 feet a week, and by the middle of May upwards of 100 feet had been executed. The beautiful machinery for lifting the excavated soil to the top of the shaft had been erected, and the men had become familiarised with their work ; but the importance of the drain, for which Brunel was induced, by the cry for economy, to substitute a pipe, now became obvious, and the want of it, as will be subsequently seen, added considerably to the expense of the work.

On the 22nd May, the top plate of No. 1 frame suddenly broke with a loud and startling report. No unusual pressure was upon it, and the only cause which could be assigned for so serious a mishap was the great change of temperature to which the top of the excavation was continually subject. This amounted sometimes to a difference of 30° Fahrenheit. Towards the end of the month (May), from want of proper supervision, the frames had become very irregular, and as the ground on the west was much more impressible than that on the east, they gradually departed from the direct line as much as 32 inches to the westward, and Brunel, finding that remonstrances were vain, determined to have the whole of the frames shifted sideways bodily, before any farther progress should be made in advance. The operation was commenced on the 4th of June, by driving a heading and timbering in the ordinary way. Fortunately, the

ground at the top proved sound, tenacious clay ; but as
the silty strata were opened beneath to the atmosphere,
large masses broke away, leaving cavities which, unless
quickly filled with clay, and closely timbered, must
have led to the most disastrous results.

By the 10th the last frame was brought over, and
the work proceeded. On the 19th, in consequence of
his having conceded to the ill-advised demand of the
directors, an attempt was made to introduce piece-
work, but the bricklayers failed to realise their wages,
in consequence of the delay caused by the bending
and breaking of the legs, in the struggle to get forward
anyhow. The clay " of sufficient depth to ensure the
safety of the intended tunnel," but found only at
the top of the working, was fast disappearing, its
place being filled by silt, which, adhering to the top
staves, produced an amount of friction in no way anticipated or provided for. This substance, although dry
and crumbly when protected, became very unmanageable when exposed to either air or water.

On the 5th of August, Mr. Armstrong was taken so
seriously ill, that upon his recovery, he resigned his situation, and an amount of labour was thrown upon young
Isambard Brunel which few confirmed constitutions
could have borne ; but all his activity could not prevent
the ground being ignorantly and sometimes recklessly
exposed, even to the extent of a whole face of a cell.

It was in this condition of the superintendence, and
when 190 feet of the tunnel had been completed, that
I was received as a volunteer on the 7th of August,
1826, and as from that period to the stoppage of the
works in 1828, and, subsequently, on their resumption
in 1835, I became personally identified with the undertaking, I shall be frequently compelled to offer the

result of my own experience as to the mode in which the operations were conducted.

The introduction of piece-work was found to be attended with most mischievous consequences. To lay the greatest number of bricks in the week was all that the bricklayers cared for; to urge on the miners their constant effort, and for six weeks the progress was 12 feet 6 inches per week. Unfortunately, for the interests of the company and his own peace, Brunel permitted the experiment to be continued. In his journal he says (August 21, 1826); " A work of this nature should not be hurried in this manner. Fewer hands, enough to produce 9 feet per week would be far better than the mode now *pursued from necessity but not from inclination*, on my part. Great risks are in our way, and we increase them by the manner the excavation is carried on. The frames are in a very bad condition." To me the want of efficient supervision became obvious, the more so where contract work was in operation, and to add to the difficulties, I soon found that there had been a considerable departure from Brunel's original design, with the view of increasing progress and satisfying the demands of the directors. More room was given to the movement of the legs, thereby increasing the angle, and causing an enormous resistance as the frames, when moved forward approached again their vertical position. The length of the poling screws was also increased—indeed, doubled—requiring from the miners, therefore, an application of physical power possessed by few. This will be readily understood by reference to the cut, page 220. The poling board at the top of a box or cell was upwards of two feet above the operator's shoulder, and, when the ground was removed from the front, upwards of a foot in advance. With his arms thus extended to the utmost, the miner was placed in the least favourable position to resist the pressure of the ground, and it ought to have been, therefore, no matter of wonder that he was frequently overpowered, and that one of the objects for which the shield was constructed should have been, in a great degree, defeated. It will be further understood that every portion of ground that had to be removed, or which was permitted to come down, above what the progress of the excavation absolutely demanded, caused disturbance in the parts contiguous, permitted cavities to be formed which became gradually enlarged, and ultimately opened the dreaded communication with the river. I cannot then withhold the opinion that had the movements been limited to $4\frac{1}{2}$ inches as originally designed, and to which they were confined during the first successful struggle, many of the difficulties which were subsequently encountered, would have been avoided, and one of the most important provisions of the shield justified.

We have seen that after mature deliberation, Brunel, in 1824, rejected the suggestion of extending the action of the frames to 18 inches; but now circumstances unfortunately combined to overrule his judgment, and to induce him to sanction what his just mechanical conceptions and his subsequent experience condemned.

The extraordinary energy, ability and enthusiasm of his son, who, upon the retirement of Mr. Armstrong, was appointed resident, seemed to offer to Brunel compensation for almost any departure from his well-considered plan. The necessity for increased supervision, however, became more and more pressing; in proof of which the following short sketch of the first serious difficulty encountered after my connection with the works will afford an example. On the morning of Friday the 8th September (1826), water was observed to drop from the tails of Nos. 7 and 8 frames. This was checked by a stuffing of oakum. In two hours diluted silt made its appearance, and during the night it burst in with considerable force, and at three o'clock in the morning, when I relieved Isambard Brunel in the superintendence, that force was so great, as to resist the united efforts of three men to retain the necessary stuffing in place. The contest was continued until six o'clock in the evening, when, the silt having been washed away, the clay settled down, and checked the flow of silt. It was now Saturday evening (9th), and the utmost vigilance was required during the whole night to retain the men at their posts, that the works might be secured against the usual suspension of the work on Sunday (10th); nor until eleven o'clock on Sunday morning, with the exception of a feverish dose of three hours, were we enabled to retire to rest. On Monday (11th) the contest had again to be renewed ; water and silt occasionally bursting from the back of No. 6 frame when any attempt was made to move on, and Mr. Brunel senior, who constantly supported our exertions with his presence, being unwilling to disturb the top staves, directed that they should be detached from the head. Timbers were introduced in front, where the ground was more solid, and capped with clay were forced up by powerful screw-jacks. While this operation was going on in front, gravel and broken pieces of yellow mottled clay forced themselves in behind. Upon an effort being made to move forward the contiguous frames, water appeared in front in such abundance, as to threaten destruction to the faces. To relieve the ground, borings were made through the brickwork of the centre pier, and pipes inserted at the back of Nos. 6 and 7 frames. After considerable labour, at ten o'clock that night (11th) the object was attained, and the water flowed with great velocity, promising to relieve the pressure, and to prevent the farther dilution of the silt and clay. All was now in full activity; the din of workmen and the plashing of the water, broken in its descent of 22 feet by the iron floor plates, was deafening, when suddenly the water ceased to flow. The workmen ceased their labour; not a sound relieved the intensity of the silence. We gazed on one another with a feeling not to be described. On every countenance astonishment, awe perhaps, was depicted, but not fear. I saw that each man, with his eyes upon Isambard Brunel, stood firmly prepared to execute the orders he should receive with resolution and intrepidity. In a few moments — moments like hours — a rumbling, gurgling sound was heard above; the water resumed its course; the awful stillness was broken; life and activity once more prevailed; and the works proceeded without farther material interruption. Shortly after the eventful pause which I have endeavoured to describe, what was my amazement, upon visiting the bottom boxes, to find men fast asleep, with the water within a few inches of their heads. By five o'clock on Tuesday morning (12th), all alarm having subsided, we ventured to retire to rest. It will thus be understood how important it had become that a regular staff of efficient assistants should be provided; for, however admirable the conduct of the men — and it was beyond all praise — yet more than one proof was subsequently given, that without the guiding and controlling power of an officer, many would have abandoned their post. The first difficulty, commencing on the 8th, had required fifty-three hours of almost consecutive attendance; the second, twenty hours. Nor were these isolated cases. My journal records twenty-four hours on the following Thursday and Friday; on Saturday and Sunday twenty-one hours. And the demands upon Isambard Brunel were still more onerous; for, with the exception of a few hours, he never left the works until the 13th, only taking sleep by snatches on the stage. But the promptitude and courage exhibited by him were not confined to underground operations. A few days previous to the first contest with the soil, the feed-pipe of one of the boilers of the steam engine burst. To stop the pumps might have been attended with considerable inconvenience, if not danger. As it chanced, Isambard and I were on the top of the shaft. Alive to every unusual sound, we ran to the engine-house. Isambard at once perceived the nature of the accident. Seizing some packing and a piece of quartering (timber four inches square), he jumped upon the boiler, applied the packing to the fissure, and one end of the quartering upon that, jamming the other end against the slanting roof of the building; but finding that the roof was being raised, he clasped the quartering, and there hung, like the weight on the safety-valve, until I was able to procure sufficient weight to attach to the timber, and relieve him from his perilous situation. By this expedient time was gained, the other boiler was filled, and the steam-engine continued its uninterrupted work.

The increasing demands which were being made upon his son's physical power, determined Brunel to ask the directors for permission to engage competent assistants, and, in justification, laid before the Board a written detailed report of the events of twenty days, of which a few of the more prominent incidents have been related, and which he had already communicated verbally to the Board. Much to his surprise, he found himself charged by the chairman with an unworthy attempt to mislead the directors, and it was stated that those extreme difficulties were for the first time brought to their notice. It is only justice to state that the opinion expressed by the chairman was not participated in by the Board, and that Brunel's demand was complied with by the appointment of Mr. Gravatt, son of Colonel Gravatt, of the Royal Engineers, who had been early destined for the profession of the civil engineer, having received an enlarged scientific education from his father, and having just completed a course of instruction in practical mechanics in the factory of M. Donkin.

With Mr. Gravatt was subsequently (in November) joined Mr. Riley, a young gentleman also expressly educated for the profession, but unfortunately with a constitution scarcely fitted to meet the demands to be made upon it.

On the 22nd October, Isambard Brunel was taken ill. On the 29th I was confined to my bed. For ten days neither of us was in a condition to resume active superintendence. This entirely devolved upon Mr. Brunel himself and Mr. Gravatt, the one or the other frequently remaining all night in the frames, and Mr. Gravatt more than once thirty-eight hours.

By the end of the year the legs, heads, and the top staves of nearly all the frames had been renewed. Many of the operations connected with the removal of

the top staves and heads were not only attended with
danger, but all involved considerable delay ; as a consequence, the progress in sixteen weeks little exceeded
an average of 7 feet. The great and continuous obstruction was the want of a drain, for which hand-
pumps furnished a very imperfect substitute. The
valves were continually being choked and the leather
cut. The bottom of the excavation being therefore
alternately full of water and empty, though never dry,
left no alternative but to carry the work on in water,
while the expense was increased 150*l.* per week. To
all this was added a moral evil for which we were little
prepared. A class of men were brought into close
connection with the working, whose habits were foreign
to such operations ; and, however characteristic may
be the daring of Irish labourers when under the
happier influences of an open sky and free air and
the applauding voices of their comrades, there was
something, in the circumstances by which they were
surrounded in the tunnel, so new and incomprehensible, that their energies seemed entirely paralysed, except
for flight. Any unusual activity amongst the miners —
any sudden gush of sand, or rattling of gravel upon the
frames, would drive them precipitately from their post.

An incident which occurred while the shield was
being restored, after the first irruption, will illustrate
the effect of panic even upon men, who, under ordinary
circumstances, could act with firmness and courage.
Mr. Brunel, transcribing from his son's journal, records the following : — " At two o'clock in the morning of the 17th October (1827), Kemble, the over-
ground watchman, came stupefied with fright to tell me
that the water was in again. I could not believe him
— he asserted that it was up the shaft when he came.

This being something like positive, I ran without my coat as fast as I could, giving a double knock at Gravatt's door in my way. I saw the men on the top, and heard them calling earnestly (down the shaft) to those who they fancied had not had time to escape; nay, Miles had already in his zeal thrown a long rope, swinging it about, calling to the unfortunate sufferers to lay hold of it, encouraging those who could not find it to swim to one of the landings. I instantly flew down the stairs. The shaft was completely dark. I expected at every step to splash into the water. Before I was really aware of the distance I had run, I found myself in the frames in the east arch. Nothing whatever was the matter, but a small run in No. 1 top, where I found Huggins (foreman) and the *corps d'élite*, who were not even aware that any one had left the frames."

As some of my readers may prefer to have the evidence placed before them in the quaint phraseology of one of those who had permitted himself to be overpowered by the sympathetic influence which panic is found to exercise, I venture to transcribe in a note the statement as it appears recorded in my journal.*

* " I seed," said Miles, " them there Hirishers a come a tumbling thro' one of them small harches like mad bulls — as if the d——l kicked 'um — screach of Murther ! Murther ! Run for your lives ! Out the lights !¹ was the d——l, like a cart-load o' bricks shot on my head — my ears got a singing, Sir — all the world like when you and me were down in that 'ere diving-bell — till I thought as the water was close upon me. Run legs or perish body, says I ! when I see

¹ This exclamation " *Out the lights !* " was more than once used by the Irish labourers, when under panic, and seemed to arise from a confused notion that the Thames was a living thing, which, being deprived of light, could not find its way so readily into the works and thus allow them time to effect their escape.

The necessity for the employment of pumpers at the shield, where, from two in August, 1826, they had, in February, 1827, increased to the number of forty, was a serious evil, and one from which we should have been free had a drain, as originally contemplated, been provided. It has been matter of condemnation that Brunel had not sunk his shaft considerably lower, and thus have secured a greater depth of ground between the tunnel and the river; but had such an attempt been made it must have failed; every frame would have been engulfed in a quick-sand, which was known to the geologists as the forty-feet sand, and of the presence of which there was alarming evidence. On one occasion, had not the eastern frame been slung to the contiguous frames, and long timbers laid under the shoes, the whole would have gone down, as a five-feet crowbar and a ten-feet rod disappeared from my hands one anxious night, when the hard crust of gravel which covered the sand had been broken through.

Indeed, to consolidate the bottom, many expedients were resorted to—rubble stones rammed down—short piles driven in; but as these means involved a disturbance of the crust, they were quickly abandoned, and beech planks substituted with excellent effect.

I have said that " blue clay of a sufficient depth to insure the safety of the intended tunnel," which was so confidently promised, proved in reality a myth. A

Pascoe a-head o' them there miners coming along as if the d——l was looking for him. Not the first, my lad, says I, and away with me — and never stopped till I got landed fair above ground. Then I began bellowing like mad for the rascals to get ropes and throw 'um down, making sure the water was coming up the shaft. Well, Sir, we was a swinging about the ropes, but the d——l a one would lay hold. So I look'd down and what should I see? why nothing at all, Sir! All a hoax!"

variety of heterogeneous strata occupied its place, in one part amounting to nine in six feet. At the line of the junction of these strata water frequently appeared. These filaments of water, as the lowest level was approached, were found to be very sensibly acted upon by the movements of the tide. This will be understood by considering the bed of the Thames as a great basin, with the different strata cropping out at different distances down the river — the lowest stratum of green gravel having its outcrop the farthest down, and the other strata following in succession ; hence, it was always observed that, as the tide rose, that part of the excavation which first felt its influence was the green gravel at the bottom. The head of water, constantly increasing, pressed upon the natural outlet of the springs, forcing the water back into the tunnel, the point of least resistance. Had this deduction been acted upon, and the ground only opened during the tide of ebb, the most formidable of the difficulties would have been avoided. But progress — progress — must be made with as much security as an energetic, but exhausted, superintendence could maintain.

On the 27th of January, 1827, Mr. Riley, the second assistant, was taken ill—fever supervened. On the 5th of February he became delirious, and on the 8th he was dead ; cut off at the early age of twenty-four. —Mr. Brunel's journal records, " Isambard, Gravatt, Beamish ill—Munday and Lane, foremen, very ill." I may add, that on the evening of the 10th, whilst superintending the operations of the bottom boxes, and seeing that all was made safe for Sunday's rest, a peculiar and indescribable sensation came over me—a haze rose before my eyes, and, in the course of half an hour, I had lost the sight of my left eye. The active treatment to which I was subjected, while it prevented me from resuming my duties until the 7th of March, proved only partially successful in restoring the vision.

So accustomed had the miners become to the movements of the ground, that small slips were little regarded. The progress had increased to an average of 13 feet 4 inches per week; 3 feet having been actually accomplished in one day. The number of men employed had increased from 180 in October to 467 in March. This included the extra labourers required to remove the increased quantity of excavated soil, and pumpers. Of the regular miners and bricklayers seven per cent. were on the sick list.

On the 26th February (1827) permission was given by the directors that strangers might visit the works upon the payment of one shilling, and proceed down the western archway about 300 feet. This resolution, though calculated to produce some little return to the company, tended to increase our anxiety. And although the men were become more familiar with their work, and better prepared to meet the difficulties from which the operations were never entirely free, yet to place the public in a position to participate in a panic was to incur no small risk. Progress was, however, being made, but the ground showed everywhere weakness. So diluted had the silt become, that early in April the miners in the top boxes had, on removing a poling, only to scrape away the stuff with their hands, while a man attended behind to supply clay, which they rammed in and pressed against the flowing ground with the aid of the poling screws. Notwithstanding every effort, whole faces would come down, the diluted silt oozing out, and leaving a cavity in front of the polings, which, having no pressure against them, fell.

R

Early in March the faces were in such a condition " as not to be trusted," and orders were given to have straw, clay and oakum at hand in the top boxes.   Four six-inch pumps were at work in the back of the west frame, and a double pump in the east, together with a horizontal pump which was worked by the engine.

And now the want of an efficient drain was felt in all its magnitude.   In the west corner so little of original ground had been left, that on the 21st of April (1827) stones, brickbats, pieces of coal, bones, and fragments of glass and china came down into the frames.   After many applications from Brunel to the directors, the hire of a diving-bell was conceded ; with the utmost alacrity and kindness, Captain Parish, in authority at the West India Docks, despatched the one used by him.   An examination was immediately made of the bed of the river, over the west corner (No. 1 frame), when, as was anticipated, a considerable depression was found, and the ground proved so loose that an iron rod could be pushed down to the frame. Subsequently an iron pipe having been passed through the ground, a direct communication was held with the diving-bell, and some gold pins supplied by Mr. Benjamin Hawes, junr. (now Sir Benjamin), were passed up the tube to be presented to friends as a memento of this extraordinary communication.   Meantime a steening was prepared, and, by the exertions of Isambard Brunel and Mr. Gravatt, was sunk from the bell over the western frame, and, the water being removed, it was filled with concrete, and proved to some extent effective.   Some of the old watermen stated that about that part of the river we shouldn't find much gravel, " it were all dredged away for ballast, and some anchors must have got pretty well into the clay."   I may mention that a shovel had been left behind near the steening at the bottom of the river ; the manner of its recovery twenty days afterwards was something ominous.

May-day dawned upon us with little of the beauty and less of the hope with which it is invested by the poet. 540 feet of tunnel had been completed. The difficulties had increased, and those in whom we had relied, taking advantage of our necessities, not only struck for increase of wages themselves (though earning from 3s. 3d. to 3s. 9d. a day), but used every effort to prevent others from entering the works. Nothing remained but to make all secure with the aid of the foreman and some of the most respectable of the men, who would not lend themselves to what they looked upon as a disgraceful proceeding.

Finding that nothing was to be gained, the disaffected expressed their willingness to return ; we were only too glad to receive all but the ringleaders, whom Brunel peremptorily refused to admit. The strike proved more serious than had been contemplated. The ground had settled down, and had become more diluted. The sliding plates and screws had rusted, and offered great resistance, while the shattered condition of some of the frames required an unusual application of power to put them forward, which demanded the most active superintendence, extending sometimes to twenty-seven hours without intermission.

On the 11th of May a sheaf of a block was found in the ground, which had come in behind No. 5 frame, and a piece of brass and an old shoe buckle in that which came down behind No. 9 ; and on the following day (12th) the shovel, which had been left behind at the bottom of the river on the 2nd of April, made its appearance with the ground in No. 6, and was perfectly identified. On the same day, in attempting to work down the top face of this frame, the ground came in with prodigious force, not now diluted, but in masses. The top box was filled and cleared, and again filled, when it was necessary to timber the back and sides of the box, and there leave it. The yellow mottled clay which overlaid the silt was broken, and, being subjected to the whole pressure of a column of water of about thirty feet at high water, would force through apertures into which scarcely two fingers could pass, when it would swell out to the size of a man's head. At length, on the 12th of May, the resident engineer, finding that it was scarcely possible for the best men to prevent the face of the ground falling in when the top polings were removed, directed that the movement of the frames should be limited to seven inches. But the mischief had been already done.

If I have been led to enter into too great detail in the following narrative, and if I refer too exclusively to my own journal, it is because the catastrophe to which the narrative relates was the first of those that tested to the utmost the mechanical value of the shield, and the strength of the tunnel itself; and because the events cannot have been elsewhere recorded from personal experience.

At two o'clock on the morning of the 18th of May I relieved Isambard Brunel in the superintendence of the working. At five o'clock, as the tide rose, the ground seemed as though it were alive. Between Nos. 6 and 7 frames there were occasional bursts of diluted silt, which subsided, however, as the tide ebbed. The men who came on at six a.m. exhibited extreme reluctance to go to their work, hanging about the engine-house, where arrangements had been made for them to dry their clothes. The day passed on with not more than the usual amount of alarm ; but, as the flood-tide returned, the same general disturbance of the ground was observed as had occurred in the morning.

The visit of a dear friend (Lady Raffles) with a large party, about five o'clock p.m., did not tend to allay a strong feeling of apprehension which took possession of my mind. No sooner had she taken leave than I prepared myself for what, I was satisfied, would prove a trying night. My holiday coat was exchanged for a strong waterproof, the polished Wellingtons for greased mud boots, and the shining beaver for a large-brimmed south-wester.

The tide was now rising fast. On entering the frames, Nos. 9 and 11 were about to be worked down. Already had the top polings of No. 11 been removed, when the miner Goodwin, a powerful and experienced man, called for help. For him to have required help was sufficient to indicate danger. I immediately directed an equally powerful man, Rogers, in No. 9 to go to Goodwin's assistance ; but before he had time to obey the order, there poured in such an overwhelming flood of slush and water, that they were both driven out; and a bricklayer (Corps) who had also answered to the call for help, was literally rolled over on to the stage behind the frames, as though he had come through a mill sluice, and would have been hurled to the ground, if I had not fortunately arrested his progress. I then made an effort to re-enter the frames, calling upon the miners to follow ; but I was only answered by a roar of water, which long continued to resound in my ears. Finding that no gravel appeared, I saw that the case was hopeless. To get all the men out of the shield was now my anxiety.

This accomplished, I stood for a moment on the stage, unwilling to fly, yet incapable to resist the torrent which momentarily increased in magnitude and velocity, till Rogers, who alone remained, kindly drew me by the arm, and, pointing to the rising water beneath, showed only too plainly the folly of delay. Then ordering Rogers to the ladder, I slowly followed.

As a singular coincidence, I may here remark that this man, Rogers, who showed such kindly feeling and devotion, had served with me in the Coldstream Guards.

As I descended from the stage, the water had so risen in the tunnel, that all the loose timber near the frames, the cement casks, and the large boxes used for mixing the cement, were not only afloat, but in considerable agitation. The light was but barely sufficient to allow me to grope a way through these obstructions, which, striking against my legs, threatened seriously to arrest my progress. I felt that a false step could not be retrieved, clad as I was, and with heavy boots quite full of water. After a short struggle, I succeeded in gaining the west arch, which, having been appropriated to visitors, was comparatively free. The water was perceptibly rising; it had already reached my waist; still I could not venture to run, feeling that a stumble might still prove fatal. If I could only gain the barrier which limited the ingress of visitors, I should be clear of the floating timber which must be there arrested! As I approached this barrier, the sight of some of our most valued hands cheered me. Not understanding the cause of procrastination, they could not withhold their expressions of impatience, Mayo and Bertram swearing lustily at my apparent tardiness. Arrived at the barrier, four powerful hands seized me,

and in a moment placed me on the other side. On
we now sped. At the bottom of the shaft we met
Isambard Brunel and Mr. Gravatt. We turned. The
spectacle which presented itself will not readily be forgotten. The water came on in a great wave, everything on its surface becoming the more distinctly visible
as the light from the gas-lamps was more strongly
reflected. Presently a loud crash was heard. A small
office, which had been erected under the arch, about
a hundred feet from the frames, had burst. The
pent air rushed out; the lights were suddenly extinguished, and the noble work, which only a few
short hours before had commanded the homage of an
admiring public, was consigned to darkness and solitude. It only remained to ascend the shaft, but this
was not so easy. The men filled the staircase; being
themselves out of danger, they entirely forgot the
situation of their comrades below. For the first time
I now felt something like fear, as I dreaded the recoil
of the wave from the circular wall of the shaft, which,
if it had caught us, would inevitably have swept us
back under the arch. With the utmost difficulty the
lowest flight of steps was cleared, when, as I had
apprehended, the recoil came, and the water surged
just under our feet. The men now hurried up the
stairs, and though nearly exhausted, I was enabled to
reach the top, where a new cause of anxiety awaited
us. A hundred voices shouted, a rope! a rope! save
him! save him! How any one could have been left
behind puzzled and pained me sorely. That some one
was in the water was certain. With that promptitude
which ever distinguished Isambard Brunel, he did not
hesitate a moment. Seizing a rope, and followed by
Mr. Gravatt, he slid down one of the iron ties of the

shaft. The rope was quickly passed round the waist of the struggler, who proved to be old Tillett, the engine man. He had gone to the bottom of the shaft to look after the pumps, and being caught by the water was forced to the surface, from which he would speedily have disappeared, but for the presence of mind and chivalrous spirit of his officers.

The roll was now called, when, to our unspeakable joy, every man answered to his name; and we were thus relieved from the painful retrospect that must have followed any sacrifice of life.

To convey the intelligence of the disaster to Mr. Brunel was our next object. Taking a pony belonging to Isambard, I hastened to town; but Mr. Brunel having gone to dine with a friend, it was not until ten o'clock at night that I was enabled to execute my painful mission.

After the first shock, and upon receiving the assurance that no life had been lost, it was marvellous to witness the elasticity with which he took part in the preparation of a letter to the " Times," which should convey to the public correct information as to the nature of the catastrophe, and in determining the mode by which he proposed to overcome the effects of the misfortune.

At one o'clock I threw myself on my bed; and, breathing forth a grateful thanksgiving for the protection which a beneficent Providence had extended to me and to those placed under my charge, I sunk, utterly exhausted, into a profound sleep, having been twenty-three hours under unusual anxiety and continuous activity.

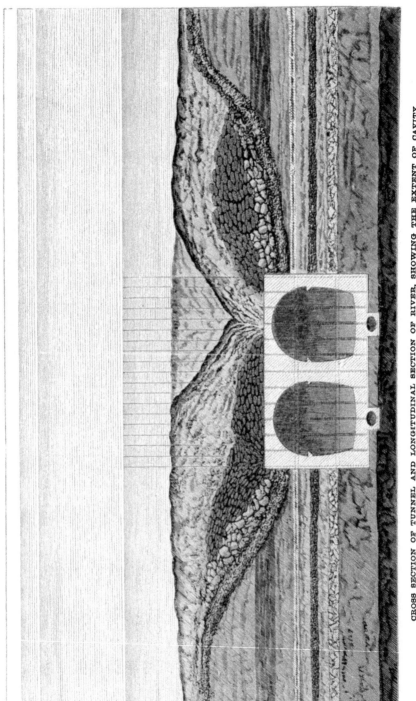

CROSS SECTION OF TUNNEL AND LONGITUDINAL SECTION OF RIVER, SHOWING THE EXTENT OF CAVITY FORMED BY THE IRRUPTION.

CHAPTER XVI.

1827–1828.

IN order to ascertain the actual state of the bed of the river, an examination was made the following day from the diving bell, when a hole was found, extending from about the centre of the tunnel excavation to a considerable distance eastward. In some parts the sides were vertical, and no gravel was found even on the contiguous undisturbed bed of the river, and confirming the observation of the waterman, that we had met a dredged hole. A tier of colliers had brought up, only the night before, over this hole, but were removed because the anchors would not hold. For fifteen years no such circumstance had occurred as a vessel attempting to anchor on this spot. And when it is remembered that harbours are sometimes cleansed by the simple expedient of mooring vessels in them, whereby the velocity of the water is so much increased beneath the vessels that the object is accomplished, there will be no difficulty in understanding how ground already disturbed would, by a small increase of the tidal action, be quickly displaced.

By a farther examination the shield was found to be in place, and by dropping almost out of the bell we were enabled to place one foot upon the back of the top staves, and the other on the brickwork of the arch (see cut). To fill the hole, and secure the work, Mr. Brunel directed that saltpetre bags should be provided. These were filled with clay, and hazel rods run through them, — the bags to prevent the clay being washed away, and the rods, by interlacing, to allow of their forming an arch. These were thrown into the river in such quantity, and with such effect, that on the fifth day the pumps were put to work. As a further security, a raft of timber, thirty-five feet square and two feet deep, and loaded with 150 tons of clay, was, on the 30th of May, sunk with considerable difficulty, not to say danger, over the frames. In adopting this expedient Brunel acted rather from extraneous pressure than from an internal anticipation of a favourable result.

The pumps were now put to work with good effect; the water was rapidly lowered in the shaft, and by the aid of a boat a visit was made, at six o'clock on the 31st, to the western arch; but at nine o'clock the pumps were overpowered, and the shaft was again filled. Upon examining the bed of the river, the western edge of the raft was found to have tilted up. A strong ebb tide had found its way beneath. An eddy was formed, and the loose ground was washed away. This result confirmed Mr. Brunel in the opinion which he had originally formed, of the inefficiency of a

OPERATIONS FROM THE DIVING BELL.

rigid covering like a raft. He therefore directed that it should be raised; but this was not an easy task in the very middle of the navigation and of a tide way with a velocity of four miles an hour.

By means of the diving bell seven chains were fixed to the eye-bolts of the raft, and on the 3rd of June the raft was raised. Before the application of clay in bags was again resorted to, a number of iron rods were laid over the opening by means of the bell; they were made to rest on the brickwork and the top of the shield, and thus to form a sort of grating. During the constant use of the diving bell, which these operations involved, many hair-breadth escapes were experienced: links in the chain, by which the bell was suspended, giving way; vessels running foul of the bell barge, and sending all adrift; and once, while Isambard Brunel was at work below, and I was superintending the signals, the foot-board of the bell, to my horror, made its appearance on the surface of the water. Although quickly reassured by the signal that all was right, yet, until my friend made his appearance, I could not divest myself of anxiety. It appeared that the man (named Pinckney) who accompanied Isambard, believing that he could stand upon the ground, incautiously let go his hold of the bell; but the ground gave way and the man only recovered his position in the bell by seizing hold of Isambard's foot, which he at once extended to him, and in the struggle the board broke away.

The application of clay and gravel was continued without intermission, and on the 11th of June, when about 19,500 cubic feet had been thrown into the hole, the pumping was effectively resumed, and a strong pressure brought upon the new ground. Mr. Brunel was again induced to give his consent to another experiment in the form of a tarpaulin spread over the new made ground. By means of anchors and chains this was accomplished, but proved as little successful as the raft.

By the 25th June the water was completely removed from the shaft, and for about 150 feet of the tunnel; but by still retaining it in the lower part, the most ready access was afforded to the shield by means of boats with ropes attached to each end; for as the tunnel inclined from two to three feet in every hundred, the water would, at the frames, be about three feet from the top of the arch.

To remove the timber, which formed an impenetrable network across the arches, was a tedious operation. On the 27th a passage was obtained for a boat, and taking with me two of the miners, Woodward and Pamphilon, in whom I could confide, we succeeded in approaching within about 120 feet of the frames. Here we encountered the ground brought in by the irruption. Leaving the men in the boat, and supplied with a bull's-eye lantern, I scrambled my way into the frames (see cut).

It would be difficult, after a lapse of thirty-five years, to re-awaken that enthusiasm which once filled the public mind; and to transcribe those expressions of admiration for the mechanical genius of Brunel which are recorded in my journal, would now appear exaggeration and rhapsody. I therefore content myself with a simple statement of the condition in which I found the shield after it had resisted the whole force of a mighty river. The frames were firmly in place, with the top staves level. The boxes or cells were about one fourth filled with silt and clay, and the bags which formed the artificial bed protruded in the front and back of the frames, as though ready to discharge their

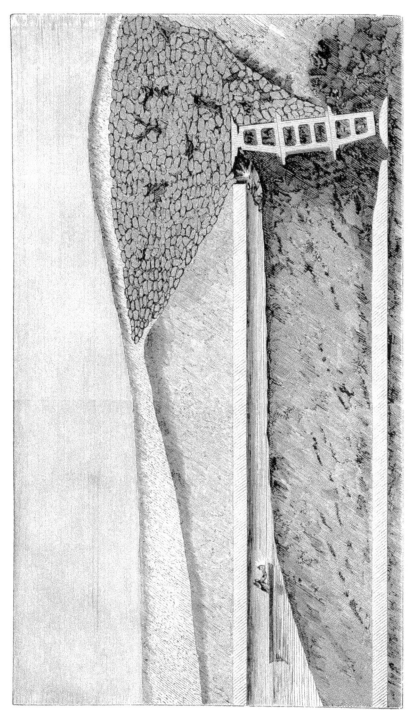

FIRST VISIT TO THE SHIELD AFTER THE FIRST IRRUPTION OF THE RIVER.

contents upon my head. The whole of the cells opposite the western arch were quite filled with silt and
clay. Water flowed in a strong stream in the east
corner. The appearance was most satisfactory, and I
hastened to communicate the result to Mr. Brunel.

The interest excited in the public mind found almost
daily expression in the public journals. It was, therefore, only natural that some of those more directly
connected with the enterprise should be desirous
to satisfy themselves as to the actual condition of
things. On the 27th June two of the directors —
Mr. Martin and Mr. Harris — having expressed a wish
to obtain a view of the shield, proceeded with Mr.
Gravatt and two miners, T. Dowling and Samuel
Richardson, in the small boat down the western
arch. Richardson, though not called upon by Mr.
Gravatt, insisted upon getting in at the stern of the
boat when he had shoved it off. The weight proved
too great. The water came in ; and though one of
the gentlemen was requested to go forward that the
boat might be balanced, neither was willing to stir,
until Mr. Martin, feeling the water inconvenient, and
forgetting that the boat had made considerable progress down the archway, diminishing rapidly the head-
way, suddenly stood up, struck his head against the
top of the arch, and fell backwards upon the others.
The boat filled, and the party were at once plunged
into water about twelve feet deep. The only swim-
mers were Mr. Gravatt and Dowling. The directors
clung to Mr. Gravatt, who could only release himself
by diving. Swimming out, he quickly returned with
a second boat. Meantime, Mr. Martin and Mr. Harris
had succeeded in getting hold of the plinth in one of
the side arches, and there supported themselves until

relieved by Mr. Gravatt; but there was no account of Richardson. A drag was procured, and his body at once found. It was conveyed to my bed as the nearest; and though every means was resorted to that science dictated, all proved in vain; the vital spark had fled — the silver cord was broken.

The shock to Mr. Brunel was great, far greater than the announcement of the irruption of the river. For that a remedy was at hand; but who could call back the departed spirit? It must be ever remembered that one of the most striking characteristics of Brunel's inventions was the means provided for the protection of life; and notwithstanding all the difficulties by which the operations of the tunnel were beset, no life had yet been sacrificed where the necessary care had been practised.

We now proceeded to secure the frames, and to clear the cells or boxes. For this purpose a cofferdam was sunk a few yards from the back of the frames; which, while it allowed of the cells being reached without waiting until the whole of the ground was removed, facilitated the operation of pumping. During the progress of securing the work, many incidents occurred, of absorbing interest to those in charge, but to recount which would now scarcely arouse the sympathy of the reader. I shall only state that after a variety of small panics, confidence was entirely restored. As an example, I give the following report made to me by the faithful Rogers. On Saturday night, July 21, a select number of men were appointed for the watch. To Rogers was assigned the weir at the entrance to the tunnel, by which the quantity of water coming from the working was measured. About three o'clock on Sunday morning he heard the foreman bricklayer (Fitzgerald) call loudly for "wedges, clay, oakum, &c.—the whole of the faces coming in— coming altogether!" Collecting what wedges and timber he could, Rogers made the best of his way to the frames, expecting to find all hurry and con- fusion, but the only sounds that met his ear were the clang of the monotonous pumping and the gurgling of water from No. 12. He examined every top box, but there was no appearance of any movement. For a moment it occurred to him that they might be all drowned, when, to his astonishment and relief, on going upon a stage erected in the west arch, there lay the careful watch, comfortably ensconced on clean straw, sound asleep. The exclamations of Fitzgerald were uttered in a dream!

During the operation of clearing and securing the frames, the ground frequently slipped in, and permitted a great augmentation in the quantity of water. Thus, on the 6th of July, the quantity suddenly increased to 1200 gallons a minute, but, on the 9th, it diminished again to 450 gallons. On that day one of the iron staples of the raft, which was sunk in the river on the 30th of May, and which had been accidentally knocked off, was found in No. 6 frame.

On the 26th of July, the frames were so far secured that the ground was permitted to be removed, and on the 28th, the oil gas apparatus was once more in operation. I may here mention that portable gas was employed during the earlier period of the working at a cost of 65s. per 1000 feet.

The harassing opposition to which Brunel continued to be exposed, and which was naturally increased by the pecuniary as well as engineering difficulties in which the tunnel operations were involved, produced, on the 11th

of August, the most serious illness to which he had
ever been subjected.

> " Nature, long oppressed, commands the mind
> To suffer with the body."

For weeks Brunel was confined to his bed or to his
room, during which time the superintendence of the
works became more and more trying, owing to the extreme foulness of the air. The ventilator in the centre
pier had been choked by the irruption, and a timber one
to supply its place had not been completed. Not only
was there a black deposit in our nostrils, and upon our
lips, but the expectoration was equally discoloured. As
a consequence, giddiness and sickness became general.
On the 14th September, I was attacked with severe
pleurisy, and it was not until the 24th of October,
that I was enabled to resume my subfluvian duties.

Many of the best men were absent from illness, and
during my absence a most valuable servant, old Green-
shield, had died. The work was proceeding slowly.

As the pressure came to be removed from the
frames, fractures were heard in all directions. On one
occasion the surge of the ground was so great as to
shake the whole of the polings. Mr. Brunel, in his
journal, September 30, says, " How slow our progress
must appear to others ; but if it is considered how much
we have had to do for righting the frames, and for repairing them—what with timbering, shoring, shifting,
and refitting, all executed in a very confined situation
—the water occasionally bursting upon us— the ground
running in like slush—it is truly terrific to be in the
midst of this scene. If to this we couple the actual
danger—magnified by the re-echoing of the pumps,
and sometimes by the report made by large pieces of
cast iron breaking—it is no exaggeration to say that

such has been the state of things. Nevertheless, my confidence in the shield is not only undiminished; it is, on the contrary, tried with its full effect."

It was not only that we had to contend against water below. The river, now, in revenge, attacked us from above. On the 1st November (1827), the tide rose three feet above Trinity high-water mark, spring tides, and nine inches higher than had been known for twenty years. The greater part of Rotherhithe was inundated; a part of the return of the coffer-dam of St. Katherine's Docks was broken down, and it was only by great activity that sufficient clay was obtained to form a bank, to prevent the water rushing down the shaft of the tunnel. A circumstance most worthy of being recorded in connection with this remarkable tide is, that the water rose and fell three times between ten and half-past eleven o'clock, a.m. First, it rose and fell nine inches; then rose fifteen inches, and fell twelve inches; lastly, rose four inches, and fell in the ordinary manner.

By Saturday, the 10th November, things were so far restored that the resident engineer determined to celebrate his success in the orthodox mode of English rejoicing, by inviting his friends to a dinner under the river. To render the effect more striking, I was requested to solicit the permission of the authorities that the band of my old regiment, the Coldstream Guards, might attend. The request was instantly granted, and on Saturday evening, the 10th November (1827), about fifty friends assembled to do honour to the occasion and to the undertaking. The side arches were hung with crimson drapery, and the tables were lighted with candelabra, supplied with portable gas. At a short distance from the bottom of the table appeared the band of the Coldstream Guards in their uniform, in accordance with the direction of the commanding officer.

As this little demonstration was to be entirely that of his son, Mr. Brunel was unwilling to appear. It was matter of course that his health — proposed by his distinguished and valued friend, Mr. Bandinel, of the Foreign Office — should be drunk with all the honours. Captain Stevens, Equerry to His Royal Highness the Duke of Gloucester, Captain Codrington, and Mr. Benjamin Hawes, junior, expressed the interest which was felt by all classes of society in the success of the great work ; and Mr. Bandinel, in proposing the health of Sir Edward Codrington, one of the earliest and most zealous supporters of the undertaking, held up a copy of the "Gazette Extraordinary" of that evening, announcing the result of the battle of Navarino. "In that battle," said Mr. Bandinel, "the Turkish power has received a severer check than it has ever suffered since Mahomed drew the sword." "Gentlemen," continued Mr. Bandinel, "it may be said that the wine-abjuring Prophet conquered by water—upon that element his successors have now been signally defeated. My motto, therefore, on this occasion, when we meet to celebrate the expulsion of the river from this spot, is — Down with water and Mahomed—wine and Codrington for ever!"

The health of the Deputy-Chairman of the Company, Mr. Wollaston, was then drunk with unmistakeable sympathy and enthusiasm.

In the adjoining arch, upwards of one hundred of the leading workmen were also provided with the means of celebrating the occasion ; and—in accordance with miners' usage — they, through their chairman,

presented to Isambard Brunel the pick-axe and spade
as symbols of their craft, over which they requested a
three times three and a bumper, which, being cheerfully accorded, the proceedings were brought to a close.
Of this entertainment Sir Benjamin Hawes, K.C.B.,
obtained a painting in oils, executed by an artist on
the spot, and which preserves, better than words, the
memory of an event which can never be repeated.

On the 20th, a general meeting of the proprietors
took place, when after reporting " that the difficulties which had arisen, and the effectual manner in
which they had been overcome, proved to demonstration the entire practicability of the undertaking,"
the directors, expressed the hope that, notwithstanding the excess of the cost beyond the estimate, " the
work would be deemed by His Majesty's Government
so far of a national character, that there would be found
a disposition to afford such farther assistance as could,
with propriety, be granted ;" and they added, " that
the interest is so general, that the success of the work
will be regarded as a national benefit and honour ; but,
its abandonment, while practicable, a national misfortune and disgrace."

The numerous repairs which the shield continued to
demand, permitted necessarily but of slow and unequal
progress during the months of October, November,
and December. As the year 1827 closed, the west
side of the excavation became seriously troublesome;
and on the 26th of December, not less than eight bar-
rows full of ground forced itself into No. 2 frame.
On the 2nd of January, 1828, some of the rocky
ground, taken from the middle boxes, and which had
been thrown into the river, came down again into
No. 2. Unfortunately, the resident engineer did not

deem it necessary to employ those means which had already proved successful, and fill up the depressions in the bed of the river. On the 8th of the month serious inconvenience was experienced in preparing for the reception of Don Miguel. He was received by Mr. Brunel and the directors, and seemed determined fully to satisfy his curiosity, by witnessing the process of excavation; nor was he contented to learn the principle of the working in the middle boxes, where all was dry, but he must ascend to the top boxes, in order, as the tide was well up, that he might there feel something of the discomfort to which the operations were liable. No change of countenance indicated that he was either surprised, gratified, or incommoded. He condescended to make one observation, and to ask one question: "That is clay"—" How far to the bed of the river?" A few days later, and he might have witnessed how small a space actually existed.

His Royal Highness was below the middle stature, presenting a very youthful appearance (though really twenty-six), dark—almost olive complexion—dark hair, and pencilled eyebrows, a full eye and tranquil mouth, but with an ill-defined nose, narrow forehead, and profoundly secretive expression.

The east stage had been moved back from the frames, that the magnitude of the work might be at once presented to the eye. On this stage, and on planks, one above another, were seated the miners. Whether from their elevated position, or the tranquil dignity of their aspect, the contrast which they presented to the eager, restless crowd below formed in my mind the most striking part of the whole exhibition. To us all, the relief was great when the last act of it had terminated.

The west corner, or No. 1. frame, continued to be a

source of anxiety, and yet no clay in bags had been applied to the bed of the river. On the morning of Saturday, the 12th of January, I came on duty at six o'clock, but was detained above ground in writing out orders for the men who had been most exposed to wet, to allow them to receive warm beer, with a little gin mixed, as had become the usual practice. I had scarcely completed the last order, when a strange confused sound of voices seemed to issue from the shaft, and immediately the watchman rushed in, exclaiming, " The water is in — the Tunnel is full!" My head felt as though it would burst—I rushed to the workmen's staircase; it was blocked up by the men ; with a crowbar I knocked in the side-door of the visitors' staircase, but I had not taken many steps down, when I received Isambard Brunel in my arms. The great wave of water had thrown him to the surface, and he was providentially preserved from the fate which had already overwhelmed his companions. " Ball! Ball!—Collins! Collins!" were the only words he could for sometime utter ; but the well-known voices answered not—they were for ever silent.

In the earnest desire to make progress, some of the precautions which experience had shown to be so important were unfortunately omitted ; and Isambard Brunel, calculating upon the tried skill, courage, and physical power of some of the men coming on in the morning shift (particularly Ball and Collins), ventured at high water, or when the tide was still rising, to open the ground at No. 1. According to his own account, given to me that day, upon the removal of the side-shoring, the ground began to swell ; in a few moments a column of ground—solid ground—about eight or ten inches diameter, forced itself in. This was immediately fol-

lowed by the overwhelming torrent. Collins was forced
out of the box, and all the unflinching efforts of Ball
to timber the back proved unavailing. So rapid was
the influx of water, that had the three not quitted the
stage immediately they must have been swept off.
A rush of air suddenly extinguished the gas lights, and
they were left to struggle in utter darkness. Scarcely
had they proceeded twenty feet from the stage than they
were thrown down by the timber now in violent agitation, for already had the water nearly reached as high as
Isambard's waist. With great difficulty he extracted his
right leg from something heavy which had fallen upon
it, and made his way into the east arch ; there he
paused for a moment to call for Ball and Collins, but,
receiving no answer, and the water continuing to rise,
he was compelled to consult his own safety by flight
Arrived at the shaft, he found the workmen's staircase,
which opened into the east arch, crowded. The morning shift had not all come down, the night shift had
not all gone up ; added to which, those who had succeeded in placing themselves out of danger, forgetful
of the situation of their less fortunate companions,
stopped and blocked up the passage. Unable to make
his way into the west arch and to the visitors' staircase,
which was quite clear, owing to the rapidity with
which the water rose, Isambard Brunel had no alternative but to abandon himself to the tremendous wave,
which, in a few seconds, bore him on its seething and
angry surface to the top of the shaft. With such
force, indeed, did the water rise, that it jumped over
the curb at the workman's entrance. Three men, who,
finding the staircase choked, endeavoured to ascend
a long ladder which lay against the shaft, were swept
under the arch by the recoil of the wave. The

ladder and the lower flight of the staircase were broken
to pieces.

We had then to mourn the loss of Ball, Collins,
Long, G. Evans, J. Cook, and Seaton. Of Ball
I can scarcely now write without feeling my deepest
sympathies awakened. His judgment, his zeal, his un-
obtrusive courage, his manly address and handsome
person, endeared him to us all. Of Collins we had had
less experience, but he also had shown himself a
valuable man.*

It again devolved upon me to make the painful
announcement to Mr. Brunel of another and more fearful catastrophe; loss of life had now to be added to
the melancholy catalogue, with a serious injury to his
beloved son. The shock could not be otherwise than
severe. One of the great advantages which he believed
the shield possessed, and one which his benevolent spirit
loved to contemplate, was the security it afforded to
life; but, unhappily, the confidence which that supposed security inspired, permitted risks to be incurred
against which no protection could avail.

Suggestions for stopping the hole, and for securing
the works against any future accident, now poured in
upon Mr. Brunel and the directors. But, as Brunel
observed, it invariably turned out that " the ground
was always made to the plan; not the plan to the
ground." Out of more than five hundred designs,
there was not one of any practical value.

Isambard Brunel was found to have received internal
injury as well as severe strain in the knee-joint, and
was confined to his bed for months.

* One of a series of drawings in possession of Sir Benjamin
Hawes, and executed by Goodall, illustrating the progress of the
tunnel, represents the recovery of the bodies of the sufferers.

In April 1828, having been called to Ireland to follow my father to his grave, I was compelled to absent myself from the works, and remained away until the 13th of May. A few lines which I received from Mr. Brunel immediately after my return, and written from his son's sick room, I venture to place on record as indicative of the affectionate interest with which he regarded our connection. "Nothing particular here, except that we feel more when distant than when near the anxieties incident to your engagement. Take care of yourself as of the shield."

During his son's illness Mr. Brunel took upon himself the duties of resident engineer. The same means were resorted to by him to stanch the hole, and to regain the frames that had proved so entirely successful after the former irruption, and about 4,500 tons of clay and gravel were absorbed by the hole. The same alarms, anxieties and fatigues were again experienced, pressing only the more heavily seeing that we were deprived of Isambard Brunel's energetic support. I may mention that a wine-glass, small basket, basin and plate, which had fallen into the river from the raft the night after the irruption, were taken from the excavation in the process of clearing—the glass uninjured.

The funds of the Company being exhausted, it was determined in July that the frames should be blocked up and the works stopped, pending an appeal to be made to the country for the means necessary to complete the undertaking. On the 5th of July, a public meeting of friends and promoters of the project was held at the Freemason's Tavern. It was attended by His Royal Highness the Duke of Cambridge, His Grace the Duke of Wellington, the Duke of Somerset, the Earls of Aberdeen and Powis, C. N. Palmer, M. P.,

W. Smith, M.P., J. Masterman, M.P., — Dundas, M.P.,
Sir John Sinclair, Sir Edward Owen, and many other
gentlemen of distinction.

On the motion of the Duke of Wellington, C. N. Palmer, Esq., M.P., was called to preside. Mr. Palmer
congratulated the meeting upon the readiness which
had been exhibited by so many noble and intelligent
individuals to uphold the undertaking as a national
work, which, if completed, would be the admiration of
the world, and a proud monument to the skill and enterprise of British industry.

His Grace the Duke of Wellington then proposed a
series of resolutions, expressing his confidence that the
great work would be crowned with success. " Of my
own knowledge," said his Grace, " I can speak of the
interest excited in foreign nations for the welfare and
success of this undertaking; they look upon it as the
greatest work of art ever contemplated. They would
not be discouraged by the accidents that had occurred.
They appeared to him, as if they had only happened to
put to the test the skill and ingenuity of the engineer.
Those accidents, however, proved that the work
already completed was sound and durable." His Grace
then read the following resolutions :—

" That this meeting being of opinion that a tunnel
under the Thames in that part of the river where the
commercial intercourse is considerable, and where
regard for the navigation must preclude the erection
of a bridge, would be a work of eminent utility; and
that the completion of such a work, would not only
be honourable to science, but would reflect credit on
the country through whose means, and continued encouragement the work was produced; and that the
undertaking at Rotherhithe, of which a large portion

is now accomplished, and which has excited at home and abroad an unusual interest, is recommended by all these characteristics ; and the abandonment (under those circumstances) of that work, which is now stopped for want of money, would be discreditable to the country : Resolved, that the prosecution of the work in question merits and demands the support of the British public.

" 2. That the present meeting being of opinion that, as the irruptions of the river had, on both occasions, been overcome by simple means, and have left the work undisturbed throughout the whole length of the part already finished, an assurance has been thereby acquired that the plan of those works is good, and is deserving of public confidence ; and, that although the enterprise was, at its beginning, of that peculiar nature as almost to baffle accuracy of estimate, yet that the completion of one-half under circumstances in which a variety of difficulties was encountered and surmounted, has given, from experience of the past, reasonable grounds for forming a conclusion as to the expense of the future ; and that it appearing, by a statement of the directors, that the premises on one side of the river, with machinery and other expenses, have cost 50,000l., and the actual progress of the work (including the accidents) has cost only 120,000l. ; making a total of 170,000l. for the whole of the present expenditure :

" Resolved, That this meeting do earnestly invite the public at large to support the plan proposed for the completion of the work, and to subscribe their names for debentures which are issuable towards it, in sums of 20l. and upwards ; and for donations, &c. ; and that books be now opened," &c.

The resolutions were carried unanimously ; and in

proposing a vote of thanks of the meeting to the Duke
of Wellington, His Royal Highness the Duke of Cambridge expressed his hearty concurrence in the views

THE THAMES TUNNEL. WEST ARCHWAY.

taken of the undertaking, and he added, " That in the
course of the long residence which he had had abroad,

it was his pride to notice the favourable opinion which was entertained towards the final success and completion of this Thames Tunnel.   He thought the pride of Englishmen was at stake in the undertaking, and therefore called on the meeting to support it by liberal subscriptions."   The result of this meeting was, that 18,500*l.* were at once subscribed ; His Royal Highness the Duke of Cambridge, the Duke of Wellington, Mr. Maudslay, and Mr. Palmer having put their names down for 500*l.* each.

It only remains for me here to add, that the debenture scheme failed, and that the bricking in of the shield was completed early in August (1828).

Subsequently, a mirror was placed at the end of the visitor's arch.   This arch was stuccoed, and being lighted with gas, continued to be an object of great attraction for years.

CHAPTER XVII.

1829–1831.

RETURNS TO THE GENERAL PRACTICE OF HIS PROFESSION — HONOURS — CONDUCT OF THE DIRECTORS — ENTERTAIN A NEW PLAN, 1829 — THE DUKE OF WELLINGTON'S OPINION — COMPARATIVE COST OF DRIFTWAY AND THAMES TUNNEL — NEW PLAN REFERRED TO PROFESSOR BARLOW, JAMES WALKER, AND TIERNEY CLARK — OPINION, 1830 — CHAIRMAN'S CONTINUED HOSTILITY — BRUNEL RESIGNS HIS APPOINTMENT, 1831 — EFFECTS ON HIS HEALTH.

BEING now relieved from anxiety, Brunel resumed the general practice of his profession, from which his devotion to the works of the tunnel had almost entirely excluded him.

I find him immediately engaged in surveys for the Grand Junction Canal Company, and for the Oxford Canal Company; for Docks at Woolwich, and in the projection of a bridge over the Vistula at Warsaw. Besides these, he was much employed in examining mechanical projects for which patents had been secured, and in giving evidence in Courts of Law for or against their validity.

It was only natural to suppose that the position to which he had attained in the estimation of the public should excite the sympathy of his countrymen. The " cupidi novarum rerum," by which they were characterised by Cæsar, had lost nothing of its applicability.

In this amiable contest for the honour of adding the name of Brunel to the roll of their distinguished men, his native city of Rouen took the lead. In 1827 he was elected honorary member *de la Société libre du Commerce et de l'Industrie de Rouen.* M. Neel, in making the announcement, says : —

"J'ai l'honneur, Monsieur, de vous renouveller l'expression de la haute admiration de la Société pour vos talents, et de son vif désir de compter au nombre de ses membres celui dont la Normandie et la France sont glorieuses."

In January, 1828, *l'Académie Royale des Sciences, Belles Lettres, et Arts de Rouen* unanimously elected him a member of their body. In announcing to him the honour, the Secretary writes : —

"Tandis que vous attachez la gloire du génie français au plus rare monument de la Tamise en ouvrant une route nouvelle à l'art précieux que vous honorez, l'Académie de Rouen, attentive aux succès d'un compatriote, a regardé comme un devoir de son institution de fixer la mémoire de votre nom sur les bords de la Seine, qui s'enorgueillent de vous avoir vu naître," &c.   This was followed, in February, by his being elected a corresponding member of the Royal Academy of Sciences in France.   M. Charles Dupin writes : " La section de Mécanique de l'Académie Royale des Sciences ayant eu dernièrement à nommer un correspondant, c'est moi qu'elle a chargé de rendre compte des candidats qui lui ont paru dignes d'être nommés.

" Il m'avait suffi de prononcer votre nom dans la section pour réunir tous les suffrages ; et il en a été de même de l'Académie toute entière.   Vous pouvez croire à la satisfaction que j'ai éprouvée en expliquant tous les titres d'honneur d'un compatriote tel que vous."

In February 1818 he was also elected a member of *the Royal Academy of Sciences of Stockholm*, the announcement being made by Berzelius.

In August he was elected corresponding member

de la Société Royale d'Agriculture et de Commerce de Caen, and, in December, corresponding member *de la Société libre d'Emulation de Rouen.*

Mr. Law, secretary to the Royal Society of Caen, says : " Je m'empresse de vous annoncer que la Société vient d'inscrire votre nom sur la liste de ses correspondans. Elle a désiré vous donner une preuve de son estime ou plutôt de son admiration pour un homme qui a pris naissance dans notre province, et qui a porté honorablement le nom Normand dans cette île où autrefois nos ancêtres, sous la conduite du Duc Conquérant, se signalèrent par leurs hauts faits d'armes.

" Vous n'avez, sans doute, pas besoin, monsieur, du titre de membre de notre société, pour votre jouissance personnelle, et il n'ajoute rien à votre réputation ; mais il sert à vous prouver que nous avons su apprécier votre rare mérite, et qu'en vous rendant cet hommage, nous avons voulu démentir à votre égard le proverbe, *qu'on n'est point prophète dans son pays.*"

In November, 1829, he was also elected *membre de la Société française de Statistique universelle* under the Presidency of the Duc de Montmorency ; and lastly, in 1833, he received from *l'Académie de l'Industrie de Paris* a silver medal which had been awarded to him as a testimony of the high opinion entertained by the academy of the benefits which he had conferred upon practical science.

However gratifying to Brunel may have been the general acknowledgment of the high position which he had attained, he was still doomed to vexation and indignities for which I confess myself unable to account.

We have seen that the appeal made to the public for means to continue the works of the Thames Tunnel proved abortive, and that the only hope was in the aid which the Government would supply. We have also seen that the numerous schemes suggested by the kindly feeling of well-wishers throughout Europe and America, either purporting to aid the original design, or to supersede it, were rejected by the directors as impracticable in their application. Still there was one which although as little entitled to consideration as the least practicable of the four hundred and ninety nine which accompanied it, yet having been urged with much pertinacity, and having been supported by all the personal influence of the chairman of the company, was permitted to occupy an amount of time and attention which its intrinsic merit in no way justified.

The recommendations with which this plan came before the Proprietors on the 3rd of March, 1829, were:—

" That the plan was totally and essentially different from that of Brunel.

" That it was a matured plan.

" That it was accompanied by the opinion of several eminent scientific men.

" And that the price would be considerably less than one half, and probably a still smaller proportion of the antecedent cost."

To carry the plan into effect it was necessary to obtain Government aid. For this purpose the chairman waited upon His Grace the Duke of Wellington, to whom he stated that a plan had been offered to the directors, with security for the execution of it, at a price considerably below that of Mr. Brunel. To this His Grace naturally observed, that it was not the want of a plan that had caused the suspension of the works of the Tunnel, but the want of money ; and he dwelt upon the absolute necessity of security being given for the safety of the work already completed, be the plan to be adopted what it might. This security the chairman explained would be effectually provided for by an iron door to be fixed at the commencement of the drift by which the work was to be carried forward from the present termination, and which, like a valve, would entirely cut off any irruption of water and ground into the Tunnel. To this His Grace simply replied that although the water might be shut out the workmen and engineer would be shut in.

While these negotiations were pending, Lord Althorp delivered a report to the Duke of Wellington, in which he recommended that the Government should be prepared to advance 300,000*l.*, for the completion of the Tunnel according to the original design.

On the 27th of May, 1829, His Grace, in reply, stated that "his colleagues and himself thought best to postpone laying the affairs of the company before Parliament till next session, when he hoped to have the benefit of his Lordship's assistance in so doing."

Months were permitted to pass in the most frivolous negotiations. No opinion of scientific men had been produced as to the practicability of the new plan, and the directors declined to submit that plan to Mr. Brunel, who offered his assistance to enable them to form a judgment upon it, as he had already done upon the hundreds of other plans which had been submitted to the Board. At length Brunel addressed a statement to the proprietors reviewing the arguments brought forward by the directors in support of their application for power to adopt any other mode of carrying on the work. After some preliminary observations, he says :—

"Taking, then, the proceedings of the meeting altogether,—the report, the discussion and the resolution,— the arguments used by some of the supporters of this resolution were : —

T

" 1st. That my plan was unnecessarily expensive.

" 2nd. That the cost of the Tunnel when compared with what I had calculated it at, showed that my estimates were not to be relied on.

" 3rd. That it was therefore necessary to change the plan of operation, and adopt a cheaper one than that before the directors ; which, backed by contractors, under adequate security, gave a just ground to expect the completion of the work at a considerably less sum than it would cost on my plan.

" In the observations which I deem it my duty to make, I shall follow the order in which I have now stated what I understand to have been the arguments used at the meeting alluded to.

" 1st. Then, as to the expense of constructing a tunnel on my plan.

" It is well known that in the year 1808, a company was formed to effect a passage technically called a driftway, or heading, across the river at Limehouse ; it was carried to the distance of 1040 feet.

" It was *an experiment* to ascertain the cost and practicability of making a larger tunnel for general traffic, and to which it was to be the drain ; and was carried on in the old method by the best miners, and superintended by a well known practical and able engineer, Mr. Trevithick.

" Every particular of this undertaking together with the accounts of the expenditure were published.

" Now I shall proceed to make an exact comparison between the expenses incurred on that occasion and the present ; so as to obtain from two experiments on a large scale the true comparative cost of constructing a tunnel on the old system (supposing it to be possible) and on mine ; or, in other words, I shall ascertain from

CROSS SECTIONAL AREA OF DRIFTWAY 1808 REPRESENTED BY THE CENTRE FIGURE,
THAT OF THE THAMES TUNNEL 1825 BY THE WHOLE PLATE.

the published documents of the Thames Tunnel, and
the archway company of 1808 —

" 1st. What each paid for excavating and removing
a cubic yard of earth.

" Expenses incurred in carrying on the works of

	The Driftway in 1809.	The Thames Tunnel.
Dimensions . . .	5 ft. high, 2ft. 9in. wide.	22ft. high, 38ft. wide.
Total sum disbursed .	£13,319 0 11½	£175,000
Cubic yards excavated .	817⅓	22,573
Cost per yard . .	£16 6 0	£7 15 0
Deducting purchase of property and law charges, per cubic yd.	$\frac{£10,772}{817\frac{1}{3}} = £13\ 8\ 6$	$\frac{£138,224}{22,573} = £6\ 2\ 6$
Deducting brickwork .	(scarcely any used)	£107,218
Cost per cubic yard .	£12 17 6	£4 5 0

" The difference is great, but the causes are easily
explicable.

" In a heading, as in the case of the archway company, the miner has at every step to construct and fix
in place a frame work to support the ground ; and if
the soil is at all unfavourable, this operation is attended
with considerable delay and risk.

" Whereas in the Thames Tunnel this support is completely effected by means of the shield, leaving the miner
nothing to attend to but the more immediate working
of the ground ; and by the space and protection afforded to the workman, we are enabled to take full advantage of that subdivision of labour so necessary to
economy, and by the facility and certainty with which
the ground is supported without interfering with the
operations of the miners, the bricklayer, the pumper,
and the labourer can proceed simultaneously, and without impeding each other.

" The plan therefore which I have adopted, and
which till now has been held out as the cause of our
success in a work of acknowledged difficulty, is not, as
stated in the report, an apparently ingenious but really
expensive plan ; but has proved, what it was intended
to be, simply a more safe, easier, and more economical
mode of securing the ground during the excavation
than the old one.

" To pursue this comparison,— for it is not in my
power to make another,—and it would be absurd to take
ordinary canal tunnelling as a measure of the expense
and difficulty of the tunnel under the Thames,— and
passing through its loose alluvial bed, — it is worthy of
remark that we have completed and secured the work
as we proceeded, and that in spite of the difficulty we
have encountered, we have never lost a single foot of
what has been completed. We have a substantial structure, the strength of which has been proved beyond a
doubt ; whereas with regard to the driftway or heading
of 1808, nothing remains but the recollection of it.

" Dr. Hutton and Mr. Jessop,—two very high authorities who were consulted on the resumption of that work,
after the eruption of the river,—gave it as their opinion,
that although they could not pretend to say what means
might hereafter be suggested, they considered, " that
effecting a tunnel under the Thames by an underground
excavation in the old mode was impracticable."

" Yet it is to this old mode that you are now called
upon to return, under the promise of *economy, security,
and despatch.*

" With what probability I leave you to judge, as the
work with which I have made the comparison of expense, was conducted by an engineer of honour, talent,
and great experience ; and failure and abandonment of

the experiment, and the subsequent opinions of competent judges, which I have quoted, will give an idea of the practicability.

" Now with respect to expenditure merely, I beg to observe, that the company had a committee of works, and a committee of finance; and I am not aware of any proposition having proceeded from them by which any material saving was effected. Two modes were at different times suggested to me by the directors, with a view to economise our funds; one was to get the work contracted for at a fixed price; and the other was, to pay the men by piece work. With regard to contract work I have placed what observations I have to make upon it under another head. Piece work was tried; but we soon found that the work was hastily and imperfectly done, particularly when difficulties increased, and great waste of materials and time ensued in consequence. We were often obliged to do the same thing over twice, and, in consequence, it was given up. In fact, in a work like the tunnel, there must be no inducement held out to the workmen to conceal difficulties in the vain hope of avoiding them; or to hide defects in their work in a situation where inspection must be imperfect.

" As to the estimate :

" On this subject I must remark, that when weight is laid on the expense of the works under my management, the proprietors ought to have been informed at the same time of the following facts : viz. that the tunnel, subsequent to the formation of the original estimate, was increased one third in its dimensions; and the brickwork, consequently, from a rod to nearly a rod and a half per foot run.

" This alteration, made in concert with the directors

for the greater convenience of traffic, was a very material enlargement of the original plan, and necessarily induced a corresponding addition to the estimate. This was the first cause of deviation from the original estimate.

" The next was the discrepancy between the *real* state of the ground, and that expected from the local information I had gathered together, after long continued inquiry, and equally from the result from the first series of borings instituted by the company, and conducted by parties entirely unconnected with myself and uninterested in the future proceedings of the undertaking.

" Besides all this, I am free to avow that the difficulties and uncertainties of the undertaking have exceeded all anticipations.

" The estimate for the future stands on peculiar grounds ; it is an application of the actual expences of the past to the future ; it is no calculation founded on data which may prove false, but an abstract from the books of the company.

" With regard to the proposal to finish the works by contract under security :

" I object strongly to contract work for an underground situation like the tunnel. It is quite impossible to insure a sound and substantial structure. Work by contract is always done with as cheap materials and in as slight a way as can be admitted by the specification. The opinion of the committee of works, and confirmed by the directors, is already recorded in an elaborate report on this subject in March, 1827, ' in which the danger of the work being slighted, instead of being performed carefully and so as to ensure permanent durability,' is assigned, amongst other reasons, for the determination to abandon all idea of contract work.

"Contract work is but piece work on a larger scale, where the objections to the one apply, at an increasing rate, to the other; it would be, in fact, most injudicious to apply either contract work or piece work to the tunnel. They may be usefully applied where the eye can follow them; but with us, where the consequences of a failure in any part might be fatal and irremediable, and would only be discovered when too late, there must be no inducement to slight the work for either profit or speed.

"And when it is considered that the tunnel is intended to stand the wear and tear of the traffic of the metropolis, it is of the utmost importance that not a doubt should exist in the public mind as to its solidity; and it has acquired that confidence in the plan on which it has been carried on.

"With regard to security for the performance of a contract, it is evident that it should be such as to compel a contractor to finish the work at all hazards, and finish it substantially, whatever difficulties may arise to swell his expences beyond his calculation. Such securities must, moreover, cover the full value of the work done; because it is only by the completion of the remainder, that this can be rendered profitable; and it may therefore be totally lost by any serious accident occurring to the new work which might be too expensive to be worth remedying."

December had now arrived, and yet no decision had been come to by the directors; nor was it until the 30th of March, 1830, and after twelve months of negotiations, vexations, and disappointments to Brunel, that the plan which claimed to supersede that of

Brunel was referred to Professor Peter Barlow, and to Mr. James Walker and Mr. Tierney Clark, two of the leading civil engineers of the day, for an opinion as to its practicability.

The report of these gentlemen was laid before a special general meeting of the proprietors on the 22nd June (1830).

After stating that they had examined Mr. Brunel, jun., Mr. Gravatt, and several of the miners,—perused various plans, the section of borings, the journal, and other documents, and having written to Messrs. Pritchard and Hoof for specific information as to the mode in which they proposed to proceed, but without receiving any reply,—they give their " decided opinion that the plan submitted to them is by no means adapted for overcoming difficulties of the nature of those which have already been encountered, and which are likely still to present themselves. That there is therefore no probability of the tunnel being completed in the manner proposed ; and that it would, to say the least, be little better than a waste of time and of money to attempt it." And after stating that it was unnecessary to trouble the board with the reasons which had directed their judgment, the report concludes thus : " Few professional questions have ever come before us which have admitted of a more decided solution than this did, after we had been informed of the nature of the ground and of the plan proposed."

As the result of this clear and explicit opinion, a resolution was passed to the effect, that "no other plan be used than that of Mr. Brunel, and that an application be made to obtain the necessary funds from Government." On the 2nd July (1830), the proprietary were not a little surprised to learn that upon a pro-

position being made by Lord Durham in the House of
Lords to transfer to the Tunnel Company a loan then solicited by a Canadian company, the Duke of Wellington
replied, that a loan had been already offered, but was
refused by the Tunnel Company! Thus were all the
exertions of the friends of the undertaking, and the
sanguine anticipations of the proprietors, for the time,
frustrated.

Nor was this all. In despite of the very decided uncompromising opinion of such men as Professor Barlow,
Mr. Walker, and Mr. Tierney Clark, the same plan
which they had declared, after full information and deliberation, to be totally inapplicable, was, the following
year, again entertained by the directors, and a committee named to negotiate the securities for its execution. Mr. Brunel now resigned his appointment, feeling
no longer able to contend against the spirit of hostility
which had been exhibited towards him by the chairman.

" An extraordinary stiffness has come over me," he
records in his journal, "and a nervous irritability such as
I have never felt before ; " and again : " J'épreuve des
douleurs au cœur depuis quelques jours."

What to another would have proved a serious and
probably protracted illness, the excellent constitution
and elastic spirit of Brunel enabled him to overcome.

CHAPTER XVIII.

1830–1843.

FORTUNATELY the project of erecting a bridge across the Avon at Clifton, which had begun to be agitated, and upon the subject of which he received, on the 13th of February (1830), a deputation of gentlemen from Bristol, seemed quite to restore him to himself, and to act as a cordial to his bruised spirit. " I explained to them," he says, " how the lateral agitation may be prevented, and how the effects of the wind might be counteracted." At the end of the year it was determined to offer a premium for the best design. Amongst the candidates appeared the name of his son ;

and Brunel, bringing the experience of his Bourbon bridges to bear upon this new project, at once devoted his energies to render the mechanical arrangements for the Clifton bridge complete.

On the 12th January (1831) he records in his journal: " Devised this day, for Isambard's bridge, a new mode for carrying the heads of the chains, and sent him a drawing of it to Manchester."

And on the 20th : " Engaged this day on the mode of passing the chains of the bridge over the heads, with all the combinations necessary for repairing, and likewise for the compensation against dilatation."

Through the months of February, March, and April, there were few days that he was not engaged upon that interesting work, determining the details and making the drawings with his own hands.

On the 19th of March his labour, his parental solicitude, and his pride found their reward in the gratifying intelligence that Isambard had been appointed engineer to the Clifton Suspension Bridge.

The death of Dr. Wollaston came however to modify his joy. To the unswerving friendship of that distinguished man, and to that of his amiable brother, he was often indebted for sympathy and support in his hours of trial and disappointment ; and Mr. Wollaston, writing in April (1831), in reference to the shares held by his late brother in the Thames Tunnel Company, and which, with the residue of his estate, passed to his sister Anna, says, " she desired me to present them to you in testimony of her regard, as well as in remembrance of our brother's admiration of your various works."

So strong was the admiration excited on the continent by the tunnel operations, and so feeble appeared to be the efforts on the part of the company to com-

plete the undertaking, that a company in Paris, called
the Paris association, offered to purchase the interest of
the shareholders, for 250,000*l*. This offer, to the honour
of the directors, was however at once declined.

During the remainder of the year Brunel appears to
have been principally engaged on designs for docks at
Woolwich, and in perfecting those which had already
been accepted for the suspension bridge at Clifton.
When in April (1832) he had an opportunity of estimating by personal observation the commercial advantages to be derived from the bridge, he came to the conclusion that the probability of its being erected was
very remote. In his journal (April 20th) he notes
that "the trustees of Clifton bridge had a meeting:
though disposed to give their money gratis, I augur but
indifferently of such liberality. They have resolved to
draw a prospectus, and to go round with it to invite the
public to subscribe. It may fairly be inferred that the
project is sinking in public estimation. Coupling the
state of things (at Bristol) with the prospect of the
trade with the West Indies, we may pronounce at once
and unhesitatingly that the scheme of Clifton bridge is
gone by."

The final experiments with the carbonic acid gas
engine were this year completed; and which, as we
have seen, failed to realise his anticipations, — not from
want of skill, nor ingenuity to overcome the mechanical
difficulties; but from ignorance of the peculiar conditions under which the properties of the gas are developed.

The fortunate combination of iron ties with bricks
and cement in the construction of the shaft of the
Thames Tunnel suggested to him the idea of establishing
a general principle of construction which should unite

in an eminent degree utility with economy. He accordingly instituted a series of valuable experiments to see how far a cheaper material might be substituted for those materials usually adopted in the construction of arches; and also to show how far the cumbersome and expensive apparatus of centring might be dispensed with. The first series of experiments had for its object to determine the force of cohesion between ties of various character and dimension, straw, the fibres of wood, hemp, reed, laths of fir and birch, as well as hoops of iron, were embedded, in lengths of from four to five feet, in cement mixed with sand (two parts of cement and one of sand); and an apparatus being arranged to draw these substances from the cement by a force acting in the direction of their lengths, results were obtained which seemed to justify a wide application of this species of bond.

In the yard of the Thames Tunnel, Brunel proceeded to test the value of his experiments by the erection of an arch without the aid of centring.

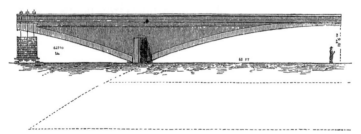

In Mr. Francis, of the firm of Francis and White, cement manufacturers, he found a ready and kind supporter, who liberally supplied the principal part of the materials and labour, free of cost to Brunel. No special preparation had been made for the foundation of the pier, indeed no ground could have been less adapted for the purpose of sustaining a weighty structure than the vegetable mould and loamy sand of the tunnel yard. The original intention was, to have constructed two semi and equilibrated arches of 25 to 30 feet span each, resting on a pier 4 ft. × 3ft. 4½ inches; but when it was deemed desirable to continue the experiment, it was found that the space would not admit of the extension of each arch beyond 40 ft. In order however to carry the principle of construction as far as circumstances would permit, one semi-arch only was advanced; while, to secure equilibrium, weights were suspended from the other. In the first part of the experiment fir ties had been employed; susbequently, hoop iron. The ultimate length obtained was *sixty feet* with an elevation or versed sine of only *ten feet six inches,* and the counterbalance required was 57,600lbs.

When subsequently the semi-arch was stuccoed, the counter weight had to be increased to 62,700*lbs.* To many, the tunnel itself was not a greater object of interest than was this remarkable structure; and though its application, in its entirety, has never been adopted in practice, yet the introduction of hoop iron bonds is now common to a variety of engineering and architectural structures.*

* Ties embedded in mortar composed of two parts of cement and one of sand: —

Materials.	Width in inches.	Thickness in inches.	Section, area.	Condition.	Strain in lbs.	Strain per square inch of surface embedded.	Results.
Fir lath	1⅛	3⁄16	·21	Half notched, half jagged.	1576		Jagged end suddenly broke.
Ditto	2	¼	·50	Notched every four inches on each side.	3093		Broke in direction of fibres.
Fir lath from saw mills.	1½	¼	·375	Left rough	2560		Drawn out.

In 1833 Brunel visited Dublin, Cork, Waterford, and Killarney, receiving everywhere the most marked attention. Upon his mind this visit left the most gratifying recollections. This relaxation, and the successes which were being rapidly achieved by his son, seemed to inspire him with a new life. It was with just pride that he saw him occupy, as it were, *per saltum*, that prominent position in the new branch of engineering from which he was himself, in consequence of the peculiarity of his engagements, excluded ; and although he saw that the public capital would be now enlisted in the more tempting speculation of Railways, he also felt that the public interest would not be the less strongly expressed in the fate of the Thames Tunnel, which had assumed the character and was regarded in the light of a national undertaking. That the government of the country should come forward, was the opinion expressed everywhere, and why it had so long hesitated to do so was incomprehensible to many. The shareholders at length discovered their error in supporting a chairman

Materials.	Width in inches	Thickness in inches.	Section, area.	Condition.	Strain in lbs.	Strain per square inch of surface embedded.	Results.
Birch hoops, area of.			·26	Pitched and sanded.	4537	31	Loaded eight days.
Iron hoops .	$1\frac{1}{16}$	$\frac{1}{16}$	·0664	Free from rust	2814	$16\frac{1}{4}$	Drawn after forty-eight hours.
Ditto .	1	$\frac{1}{16}$	·062	Faintly reddened by rust.	2387	$19\frac{1}{2}$	Drawn.
Ditto .	$\frac{5}{8}$	$\frac{1}{32}$	·0195	Faintly reddened by rust.	2055	25	Drawn.

From these experiments we find that a square inch of iron, equal to thirty-six iron hoops, though embedded only five feet in cement, would resist a force of 75,000 pounds weight, or nearly thirty-three and a half tons.

whose views were opposed to those of a majority of his
colleagues; and they, therefore, at the annual general
meeting on the 6th of March, 1832, declined to re-
appoint Mr. Smith on the Direction. A unanimity of
action being now given to their councils, the directors
were enabled to press their claims upon the country
with effect. A bill was prepared, and in the following session it was introduced to the House. Brunel
and his tunnel were again in everybody's mouth.
Royalty felt the impulse, and His Majesty William IV.
invited Brunel to give him a full detail of all the
operations. *L'Académie de l'Industrie de Paris*
awarded him, as we have seen, a silver medal for the
benefits which he had conferred upon practical science.
A Tunnel Club was established, principally from amongst
the Fellows of the Royal Society, and on the 25th of
April, 1834, the sixty-fifth anniversary of Brunel's
birthday was celebrated by the members, at the Crown
and Anchor, by a magnificent entertainment.

Curious to record, on the 29th of April, when the petition of the Tunnel Company, which had cost so much
time and labour to prepare, should have been presented
to the House of Commons, it was nowhere to be found;
and what is yet more remarkable, never was recovered.
It is not improbable but that another session might
have been permitted to pass without result, had not
Lord Althorp taken upon himself the responsibility of
granting the necessary loan from the Treasury.

Brunel, being now assured that means would be forthcoming, took the opportunity of paying a last visit to
his native country. Passing through Hacqueville, he
notes in his journal : "Je n'ai rencontré que Penchon
le menuisier de ma connaissance. Penchon doit avoir
au moins 72 ans maintenant. Il ne me reconnut point

lorsque je m'addressai à lui à sa fenêtre. J'entrai par la fenêtre à son grand étonnement, et je lui dis qui j'étais. Il me montra une partie du premier montant d'un octant que j'ai tenté, et enfin réussi de faire, ensuite quelques rouages, &c.

" Il appela sa femme, belle, il y a 50 ans, mais aujourd'hui aussi vieille que lui. Tout ébahie, elle me regardait longtemps sans parler. Enfin — Ah c'est Monsieur Isambard ! mais comme il est changé ! "

On his return home he received an application from the agent of the Viceroy of Egypt to furnish designs for effecting a secure and permanent passage across the river Nile. It appears from his journal that Brunel devoted considerable time to the examination of all the records relative to the phenomena presented by the Nile, in the British Museum and elsewhere. He found, however, that without personal observation he would be unable to fulfil the expectations of the Viceroy ; and as his engagements did not permit him to make a journey to Egypt, he was obliged to decline the commission, which was some years afterwards executed with so much ingenuity and skill by Mr. Robert Stephenson.

The Government having consented to make a loan of 246,000l. to the Tunnel Company, the first instalment of 30,000l. was advanced in December 1834, under the condition that the money " should be solely applied in carrying on the tunnel itself, and that no advance should be applied to the defraying any other expense, until that part of the undertaking which is most hazardous shall be secured." Isambard Brunel was now too deeply engaged at Bristol to permit of his resuming his tunnel labours ; and I therefore entered upon the duties of resident engineer, on the 22nd of January, 1835, with the assurance of being supplied with an efficient staff.

The four gentlemen selected as assistants by Mr. Brunel were, Mr. Lewis D. B. Gordon, Mr. Joseph Colthurst, Mr. Andrew Crawford and Mr. Thomas Page.

That these gentlemen performed their duties entirely to the satisfaction of Mr. Brunel there is ample testimony, and though it does not enter within the limits of this work to trace their subsequent career in the profession which they still continue to adorn, I cannot deny myself the opportunity to place on record the grateful sense which I have never ceased to entertain of the constancy, courage and ability exhibited by them throughout our short but eventful connection.

Had the Treasury minute been less stringent in its conditions, much time, expense and anxiety would have been saved in the completion of the work. No additional outlay would have been incurred in sinking the shaft, which ultimately had to be sunk; no expensive preparation, which, involving considerable anxiety and danger, would have been required for setting up the new shield; while the cost of removing the excavated ground would have been reduced to a minimum, and the chances of panics almost entirely removed. In short, so strongly was Mr. Brunel's mind impressed with the unreasonableness of being compelled to adhere to the letter of the minute, that he was with some difficulty prevented from abandoning the enterprise.

To show that more than doubt existed as to the practicability of removing the old shield, I may state that one of the directors declared, in my hearing, that " he would undertake to eat any part we dared remove." The difficulty was notwithstanding effectually overcome by the simple arrangements here shewn.

MODE OF SUPPORTING THE FACE AND THE TOP OF THE WORK

Not only was the ground completely supported, but ample space was allowed for the removal of the old shield, weighing eighty tons, and the introduction of the new shield, weighing 140 tons, in 9000 parts, fitted with

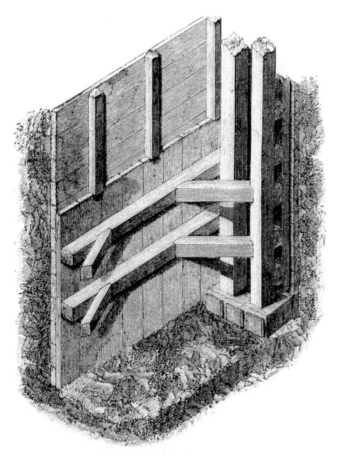

MODE OF SUPPORTING THE SIDES OF THE WORK.

all the precision required in the construction of a steam-engine.

To effect this, 1656 square feet of surface had to be supported. To relieve the pressure of the ground in front, flat iron piles, four inches broad and six feet long, and having a hole at the end, were driven into the ground in front. Into the hole a long, deep key was inserted, and the piles being driven close up so as to bring the keys against the polings, the screws which sustained the polings were capable of being relieved, and ultimately removed. Three hundred of these piles were thus employed with entire success. All these operations, however, were preceded by the construction of drains, without which it would have been almost impossible to proceed.

Reservoirs, six feet square, were formed below the inverted arches of the tunnel, and at a distance of twenty-five feet from the back of the shield. From these reservoirs drifts were run beneath the inverts to the frames, the reservoirs themselves being also united by a cross drift under the middle pier. This, though apparently a simple operation, offered considerable difficulty. From eight to ten feet beneath, the borings entered the blowing sand, to which I have before alluded. A pipe being introduced, the water rose with force to the crown of the arch of the tunnel. This determined the depth to which the reservoirs could be carried. Iron sheet piles with flanges were driven to within ten inches of the sand on three sides of the reservoirs, the drift ways occupying the fourth. This work was commenced on the 19th of March and completed on the 7th of May. Pipes from the dandy pump drew the water to the great reservoir, and the face of the working was thus completely relieved.

To the Messrs. Rennie was intrusted the construction of the new shield, in which considerable improvements were made. Slings were introduced between the frames, which, by attaching one frame to another, contributed to prevent any one from sinking when the ground in the bottom became irregular. Quadrants also were attached to the middle floor-plates for the purpose of limiting the action of the frames. Finally, the side staves were attached to the end frames by travelling pins, and each head of a frame carried two top staves in place of three.

By November the old shield had been removed, and then it was that the extent of the excavation appeared in all its magnitude. So formidable, indeed, was the cavity, that some of the directors, who were anxious to satisfy themselves that we were really prepared to receive the new shield (a fact which had been denied, and upon which the contractors for the new shield had not calculated), became so much alarmed, that after one look they turned and made the best of their way to the top of the shaft.

I may here mention that Brunel had the gratification, in the summer of this year, of seeing his son united to Miss Horsley, a connection which was productive of happiness to all parties.

By the 1st of March, 1836, the new shield was in place : the only accident that had occurred was the jamming of one man's finger, and which was the result altogether of carelessness and in spite of warning ; and it was a source of heartfelt thankfulness to Mr. Brunel, and to those immediately connected with the supervision of the works, to find the success of the first important step recognised by the directors, in their report to the proprietors on the 1st of March, 1836, where they state that they " cannot here refrain from calling the especial attention of the proprietors to the fact, that, from the first removal of the old machinery to the erection in its place of the last portion of the new shield, under, at all times, a vertical and lateral pressure of about three thousand tons, and under other circumstances of great difficulty and danger, with which the proprietors are familiar, *not only had no life been lost, but not an accident worth recording had occurred.*"

The health of all had however suffered : Mr. Colthurst was compelled to retire, and Mr. Crawford was appointed in his place.

By strictly adhering to the fundamental principle, that security should be the primary object, only slow progress was made with the excavation ; but as everything was under command, all operations were performed with increasing confidence. On the 21st June, 1836, this was tested in a remarkable manner when, during a high tide, there was so great an influx of water into the works, that the pumps were overpowered, and had not the men stood to their posts, and unflinchingly obeyed every order with the utmost promptitude and coolness, the most serious consequences might have been the result. Not less than 500 gallons of water per minute poured into the top boxes alone. The whole ground above and in front seemed to be in motion. Commencing at the west corner, it appeared to be propagated with a sweeping rushing sound over the top of the whole shield, and ultimately to concentrate its power on the east corner, until the tide began to fall, when relief was obtained ; and before another tide had risen, a quantity of clay and gravel had been thrown into the river sufficient to secure the works. Notwithstanding the enormous and unequal pressure to which the shield had been subjected, no injury had been received,—not one fracture had occurred. For some days little progress was made, the ground in the front was too loose to permit of its being opened ; but by the skill of Brunel, this difficulty was overcome. The middle and lower faces proving sufficiently sound, they were worked down, and flat iron plates or piles six inches wide were then driven vertically upwards from the front of the middle boxes beyond the polings of the top boxes till they came in contact with the top staves. These piles had the effect of cutting off a slice of ground which was subsequently removed without farther danger to the top, and in this manner a slow but safe progress was effected. By the end of August 655 feet had been secured, the centre of the river had been passed, the top boxes were almost free from water, and everything promised a rapid and safe progress ; failing health, however, did not permit me to enjoy the gratification, or secure the triumph of conducting the work to its termination.

On the 24th August, 1836, I was compelled to resign into other hands the conduct of the works. In September Mr. Gordon's health broke down, and he was also compelled to retire, the principal supervision therefore was thus thrown upon Mr. Page, who met the difficulties which subsequently presented themselves with an amount of constancy, courage and ability worthy of all praise, and which few situations have ever more rigorously or more persistently demanded.

For some months the works continued to progress favourably ; but as the old bed of the river was again entered, the ground was found to be as treacherous as heretofore and while $19\frac{3}{4}$ feet of tunnel were executed in September 1836, only $6\frac{3}{4}$ feet were with much difficulty accomplished in January 1837. This was reduced to one foot in June. At the commencement of this year it was found necessary to make application to Parliament for a farther grant of money, in consequence of the great increase of cost which the difficulty of excavation entailed.

After noticing the opinion of Mr. James Walker, C.E., who had been employed by the Government to report upon the work generally, the Committee of the House of Commons close *their* report to the House thus : —

" Looking to the importance of a work of this nature, for the first time now undertaken as a means of fixed communication to situations where no other of an equally permanent nature may be available, and also that the sum of 180,000*l.* has been already expended upon the work by the proprietors, and the farther sum of 72,000*l.* by the public, they are of opinion that it will be expedient to authorise the Treasury to continue the advances to the Thames Tunnel Company according to the Act of Parliament." Again was Brunel assured of the high opinion entertained by the country of the importance of the work : a gratification which was enhanced by the marriage just at this time of his youngest daughter to the Rev. Mr. George Harrison. Unfortunately his joy was but short-lived, for so greatly had the difficulties increased in the tunnel, that his thoughts became entirely absorbed by them.

On the 23rd of August, the river, for the third time, broke into the works, for the third time to be expelled ; but only to return on the 3rd of November following. The only resource seemed to be in the artificial bed ; which, being passed through, the same extreme looseness of the ground would again arrest all progress.   On the 21st March, 1838, the river, for the fifth time, made its way into the works ; and thus, in twenty weeks, and in a distance of twenty-six feet, no less than three irruptions of the river occurred ; but by a careful adherence to the conservative principle of the shield, one life only was sacrificed, and that not from the want of sufficient time to escape. So bad, indeed, had the ground become, that any attempt to remove a poling-board was followed by an irruption of semi-fluid matter.

Brunel therefore directed that no polings should be removed at all ; but that they should be forced forward by their screws the required distance ; thus the ground was condensed, and the miners relieved from the great difficulty of replacing the poling-boards. But there were other enemies against which Brunel was now called on to contend, and for which no previous experience had suggested to him to make provision.

The water from the springs came largely impregnated with poisonous sulphuretted hydrogen gas; the black mud which rolled in, spread its foul, noxious pestilential influence throughout the works. In vain Faraday and Taylor, and Babington and Murdock, suggested the application of disinfecting agents. A sudden irruption of sixty cubic feet of this mud and water at once neutralised all such appliances. The only mode of modifying the effect upon the health of the men, was to limit the number of hours of underground work, and to secure ventilation. Still the men gradually sunk under such overwhelming trials. Inflammation of the eyes, sickness, debility, and eruptions on the skin, were the most prevalent symptoms ; and if exertion were long continued, the men would fall senseless in the frames, often at a time when their efforts were most needed. An explosive gas, or fire-damp, also spread dismay amongst the labourers. Large puffs of fire would pass from twenty to twenty-five feet across the shield. Through all these dreadful physical trials, Mr. Page, though often compelled to

absent himself, that he might breathe for a time purer
air, continued his valuable superintendence, receiving
from time to time " the cordial thanks and approbation
of the directors, for the presence of mind and excellent
judgment which he displayed." In these well merited
acknowledgments, the services of Mr. Page's assistants,
Mr. Francis and Mr. Mason, were included.

On the 4th of April, 1840, when the works had advanced within low-water mark on the north side of the
river, and about ninety feet from the site of the proposed
shaft, the ground over the shield was observed, during
low water, gradually to sink. The utmost excitement
prevailed. People ran in crowds to the wharfs and
warehouses ; those that could obtain a boat pushed off
to the nearest vessels on the river, with the vague idea
that they might witness the destruction of the tunnel,
and perhaps the struggles of those engaged in its
execution. On an area of *thirty feet diameter*, the
ground had gone down bodily, leaving a cavity *thirteen
feet in depth*. Even by those who had the most confidence in the shield, the effect of this enormous dis-
placement was viewed with an excited, anxious and
painful feeling. What then had really occurred in
the shield ?

According to the report of the assistant on duty, a
noise was heard, which he describes as like " the
roaring of thunder ;" a rush of air immediately followed, every light was extinguished, and the men fled,
amazed and bewildered, with the exception of a few
of the veteran and experienced hands, who, although
astounded at a convulsion which threatened consequences the most calamitous, but which they were unable
to define, still preserved presence of mind, and patiently
abided the result. It was with the appearance of water

only that danger was connected in their minds, and as no water accompanied this unprecedented movement of the ground, they were unwilling to yield themselves to the panic which drove their comrades so precipitately from the works. Their example had quickly the effect of dissipating the alarm created, and the men returned to the shield. The few displaced polings were restored, and an increased sense of security took the place of that consternation and terror, which, but a few minutes before had prevailed.

The cavity formed by the extraordinary sinking of the ground was speedily filled, in the usual manner, with clay in bags, and time was permitted for the ground to become consolidated. In fact, during the two months of April and May, not more than five feet was added to the tunnel. Meantime preparations for sinking the shaft on the north shore were being made. Iron and timber curbs were laid in the manner described for the construction of the shaft on the south side. The company, from the want of funds, were compelled to limit their purchase of ground to a space contiguous to the wharf surrounded by buildings in a dilapidated condition, devoid of proper foundation, and many of them dependent upon shoring for their support. For any damage done to these buildings, which would be placed on the brink of an excavation 70 feet in depth, and 55 in breadth, the company became liable. Yielding to necessity, Brunel, on the 9th October, 1840, proceeded to clear the ground for the reception of the curbs of the shaft. Nothing could be more unfavourable than the appearance which the locality presented. On the 9th the excavation exposed a sort of timber floor, supported on piles; the piles offering great resistance, while the ground was soft and yielding, rendering it impossible to obtain an equal bearing. It soon became evident that the whole area had been at one period a ship-breaker's yard. All sorts of materials and instruments common to the ship-breaker were found in jumbled, disorganised masses : wharf pilings of two distinct periods : walings (the strong side planking of ships), masts, iron ties, bolts, chains, tools — loose ship timber, and the wreck of a boat. Yet through all these impediments Brunel continued his labour — constantly subject to the complaints of the directors for the slowness of his movements though taxing his ingenuity to overcome difficulties without parallel. The shaft in its construction differed from the former one in being sunk the whole required depth without underpinning, by which time and expense were saved ; in being given a conical form, by which friction was greatly mitigated ; and being more liberally supplied with hoop-iron ties, by which its strength was greatly increased. Notwithstanding the severe cold of December 1840, and more particularly of January 1841, which interrupted all operations, this stupendous structure, 55 feet in diameter, 70 feet in depth, and of 2000 tons weight, was executed in thirteen months — and with so much care and forethought, that not a single accident is recorded — nor had any serious disturbance of the contiguous ground taken place.

In August 1840, the excavation for the tunnel had been suspended when within sixty feet of the north shaft, and before the work was resumed a drift way was opened from the culverts beneath the invert into the shaft, thus permitting the long looked-for communication across the river to be accomplished, and at once relieving the shaft from any accumulation of water.

Before the excavation of the tunnel was resumed,

Brunel had received from the hands of his sovereign the honour of knighthood. Amongst the congratulations he most valued was one from his old and constant admirer and friend, Earl Spencer. Writing from Althorpe, 28th March, 1841, his Lordship says : " You have fairly earned your title by long continued and able services to the country ; and it is a memorial of those services, and consequently highly honourable to you. I therefore most sincerely congratulate you upon receiving it."

In July 1841, the shield resumed its functions, and by a careful and rigid adherence to the principle of its action, the ground was prevented from running in, although the influx of water was never less than 450 gallons per minute.

It is worthy of remark, that the pressure exercised by the shaft on the yielding ground being propagated to the shield, the most unlooked-for fractures were produced, leaving, ultimately, the frame on the east side a complete wreck.

On the 15th December, 1841, the last poling-board was removed from No. 1 frame. The top staves had come into contact with the brickwork of the shaft, and all that remained was to remove the brickwork, that the shield might be received, and Brunel's triumph realised. But this last, and apparently easy task, was not accomplished without its own anxiety ; not from any apprehension as to the derangement of the shaft, through which an opening of 930 square feet had to be made for the admission of the shield, but from the difficulty of completing the junction of the tunnel with the shaft. But half an inch of opening remained, yet through that half inch the semi-fluid ground forced its way, and which, if permitted to exhaust itself, would have produced settlements that would have compromised the

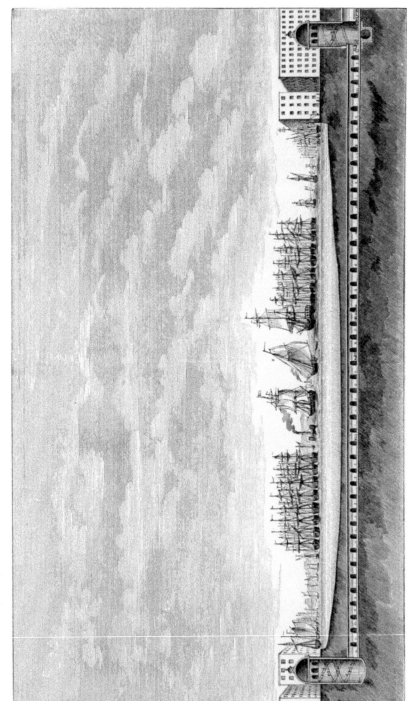

LONGITUDINAL SECTION OF THE THAMES TUNNEL.

safety of the contiguous buildings, for which the company would be held responsible. The best efforts of the best hands had, therefore, up to the last, to be called into requisition. Nor was it until the 7th January, 1842, that the work was entirely secured. To stop the waterways, and construct the staircase, occupied the remainder of the year. The influx of water, which in February amounted to 450 gallons a minute, was reduced in March to 288 ; in April to 150, and in May to 70.

Now that the battle was won the effects of that extreme tension to which Brunel's sensitive mind had so long been subjected, began to exhibit itself in the dreaded form of paralysis, although but slightly affecting his speech and features. This affliction was accepted by him as a beneficent warning, and with a calmness and resignation beautiful to contemplate. Patiently obeying, with almost child-like simplicity, the prescriptions of his medical advisers, he was enabled to ward off, in a wonderfully short time, the threatened evil, and with renovated strength to take part on the 25th March, 1843, in the ceremony of opening the tunnel to the public. " We were fearful," says Lady Hawes, " lest the excitement might prove injurious, and bring on another attack of illness ; but to our great relief he received the congratulations of his friends and the cheers of the multitude with a singular calmness, very unlike his former self; and though evidently gratified, was in no way elated."

As some evidence of the strong hold which this work retained on the public mind, it may be stated that from 6 o'clock on Saturday evening, the 26th March, to 9 o'clock on Sunday evening the 27th, or in the course of *twenty-seven hours*, not less than *fifty thousand* persons passed through the tunnel, all bearing testimony to the confidence which the structure inspired ; and though of those thousands, few perhaps could ever know or comprehend the nature of the ordeal through which those who had been instrumental in its accomplishment had had to pass, still each seemed willing to offer a tribute of admiration to the genius, the skill, the industry, the perseverance, and the courage of its gifted architect.

On the 30th of April, upwards of 495,000 persons had visited this work, and in fifteen weeks from the day on which it was opened to the public, upwards of *one million* of visitors, from almost all the civilised nations of the world, had done homage to the directing spirit, the *genius loci* of the Thames Tunnel.*

---

* The following statement shows the total Receipts and Expenditure from the commencement to the completion of the work.

RECEIPTS.	£	s.	d.	EXPENDITURE.	£	s	d.
Amount received on 3874 shares . . .	179,510	15	0	Purchase of property, rent, taxes, Parliamentary and law charges.	64,962	15	4
By subscriptions . .	1500	0	0	Machinery and labour .	338,243	16	1
Exchequer Loan Commissioners . .	250,500	0	0	Salaries to engineers, secretary, and clerks .	43,986	1	1
Rents and wharfage .	5,767	12	5	Payments to directors .	7.618	1	3
Old materials . .	3,450	14	7				
Indemnity by loss from fire	40	0	0		454,810	13	9
Interest on premium on Exchequer Bills .	3,083	16	0	To pay interest on Exchequer Bills, 31st December, 1844 .	13,439	3	7
Visitors to view the Tunnel and sale of books to 31st December, 1844 . . .	24,396	19	4				
	£468,249	17	4		£468,249	17	4

CHAPTER XIX.

THE necessity of consecutively tracing the public
career of Brunel, and of exemplifying the nature
of his works, having left but little opportunity for
illustrating his personal, social, and domestic qualities,
I have thought it advisable to devote a special chapter
to those objects.

Brunel was below the middle stature, his head con-
spicuously large, though without destroying the symmetry of his person ; so striking, indeed, was his
forehead, that an Irish friend of mine, after his first
introduction, was tempted to exclaim, " Why, my dear
fellow, that man's face is all head ! "

His temperament was nervous, sanguine, lymphatic. The anterior and posterior lobes of his brain were unusually developed. The union between the perceptive and reflective faculties was, with the exception of his gifted son, more complete than I have ever met, and, to adopt the nomenclature of the phrenologist, I may state that of the moral sentiments, benevolence, veneration and hope, largely predominated. Of the selfish sentiments, love of approbation exercised the greater influence; and of the protective, secretiveness. Brunel's joints and muscles were singularly flexible. Lady Hawes mentions two amusing examples of the use to which he turned this power. Upon his first arrival at Falmouth, he sent for a tailor to take his measure for a coat. He was in haste, he said, and the tailor promised to have the coat ready to be tried on the following day, True to his promise he appeared. The coat was tried; but the *right* shoulder was discovered to be so much higher than the left, that to fit it was impossible. The poor man offered many apologies for not having observed the peculiarity, and promised to lose no time in correcting his error. In the evening he returned, when, to his dismay, he found the peculiarity of formation was in the *left* shoulder, not in the right. He was perplexed, distracted, and utterly at a loss to account for his blunder. The coat would be useless. Time and money gone! At length, when in a climax of despair, Brunel undeceived him, and, promising to compensate him for his trouble, the poor tailor joined heartily in applauding the joke.

On another occasion a woman claimed his charity, which was indeed seldom withheld; and almost forcing herself into the house, piteously bewailed the accident that had entirely deprived her of the use of her thumb, and therefore of her needle,—nothing less, in short, than severe dislocation. The manner of the woman, however, aroused Brunel's suspicion. He examined the thumb. " Ah !" said he, "that is very curious ! Almost as curious as my thumbs," and, to the woman's amazement, he exhibited both of his, precisely in the same condition as she pretended hers to be. He had discovered the cheat, and the impostor instantly quitted the house.

His habits were in the highest degree simple and unostentatious. · Instinctively conscious that no greatness can be obtained without disinterestedness, he followed the natural impulses of his nature with a constancy and devotion from which no physical gratification could ever seduce him. When I was first favoured with his friendship, all he seemed to desire was leave to pursue the conceptions of his capacious and active mind in the quiet retirement of his study ; his drawing-board before him, and his ready hand giving a practical direction to the fertile suggestions of his teeming brain. Not that he was insensible to the charms of social intercourse : on the contrary, never perhaps did he appear so amiable, as in the society of educated women ; for in no one was there a stronger appreciation of that grace, dignity and refinement, which distinguish the higher ranks of female society in this country.

His religious sentiments were hopeful and sincere. Although educated as a Roman Catholic, his mind was of far too expansive a nature to be " cabined, cribbed and confined," by forms which had usurped the place of devotion, and ceremonies which had lost their symbolical significance.

From the time of his landing in America, he never

entered a Roman Catholic Church with a devotional object; and in a letter to his friend Mr. Ellicombe (June 20th, 1816), he says "Could I believe what I saw on the stone on the borders of the Meuse, I might be tempted to try the charm of the oraison. Unhappily for me, my faith in such things is petrified, otherwise I should try the benefit of the invocation." A sketch of the stone was subjoined with the inscription : —

" En disant trois Ave Maria, on gagne quarante jours d'indulgence."

During his long sojourn in this country, he was satisfied to adopt the forms of the Church of England, to which Mrs. Brunel was strongly attached. His knowledge of the Scriptures was extensive, if not profound ; and certainly it was to the astonishment of many, that a Frenchman should be found so well read, and so unexpectedly prepared to vindicate his theological opinions. We have seen how strongly his mind was impressed with the value of constitutional government. For in no one indeed did law and order, and intellectual progress, form more entirely the true elements of social life.

In society he was a great favourite, as well from the variety and accuracy of his knowledge, as from a naïveté and humour of expression, which was much enhanced by his foreign accent ; and though not unwilling to enter into new topics of conversation, his natural disposition led him rather to indulge in anecdotes of the past.

The serious and the gay often trod closely upon one another, in the mind of Brunel. Writing to his friend Mr. Ellicombe, under a deep sense of vexation and disappointment at Mr. Ellicombe's sudden and dis-

courteous dismissal from Chatham by the Navy Board,
he says : " Continue to trudge on this stony road until
you reach some resting-place, which, like the stone I
once saw on the barren mountains in Scotland, invited
by its inscription the traveller to ' *Rest and be thankful.*' But what think you ? this kind memento did
frighten so much a tourist of my acquaintance, that he
turned about, and measured back his steps all the way
to the capital — 160 miles from it ! "

And when a few years after Mr. Ellicombe was
united to Miss Nicholson, Brunel writes : —

" It is a lucky thing that you are born within the
pale of the Protestant Church, for had your star cast
you under the Roman hierarchy, no such wife would
you have been allowed to take to yourself, however
virtuous, however exemplary. Your present partner
even would have been excluded ! "

If a too anxious desire for the good opinion of others
sometimes robbed Brunel of that dignity which the
English mind desires to attach to high mental development ; or if it sometimes excited a feeling of super-
cilious contempt in those whose intellectual or moral
powers were far inferior to his, it rendered his character more amiable, more lovable, and more tolerant,
than if it had been under the dominion of that most
uncompromising of all the motive faculties—self-
esteem.

It cannot, however, be denied that this sensitiveness
to the opinion of others was often productive of pain,
and sometimes of an irrepressible expression of mortification. When engaged upon the works at Chatham,
Mr. Seppings (afterwards Sir Robert) ventured, as we
have seen, to ridicule his fears as to the stability of a
large chimney, which was then being erected—fears

which proved to be only too well founded. For a long time the ridicule of Mr. Seppings produced a soreness and irritability, the expression of which was confined to his journal; at length he could endure the recollection of what he looked upon as an indignity no longer, and could not resist addressing Mr. Seppings upon the subject. A protracted correspondence ensued; which Seppings was obliged to close thus :

"I trust that you will desist from again addressing me on this subject, it being unconnected with my profession : at the same time I am ready to give you every assistance in my power on this and every other occasion, you still having my best wishes.— I remain, dear Sir, yours most truly,

"R. SEPPINGS."

When in 1821 he was made the victim of the imprudence, mismanagement, and dishonesty of others, and was actually imprisoned in the King's Bench, this "exquisite sensibility of contempt" seemed utterly to have overwhelmed him. He writes to Lord Spencer July 24th:—"I have now been in this distressed situation ten weeks. I summoned as much fortitude as possible to support the misfortune ; but I find I can no longer bear up against what in the eyes *of the world* must appear a disgrace."

Such acute sensibility to the opinion of others is most commonly found in connection with affectation and pretension. Not so with Brunel ; modesty and simplicity characterised his intercourse, and gave to his society a peculiar charm. Lady Hawes mentions his having been once invited to dine with His Royal Highness the Duke of Sussex at Kensington. He did not make his appearance until the company were already seated at table. With the utmost ingenuousness he excused himself, saying, that "it was the fault of the *omniboos* that *would* not bring him quicker." To the love of children we instinctively attach simplicity, ingenuousness and purity. In Brunel these qualities shone with a constant and steady light. At Rotherhithe his study window opened to a court where young life abounded. Into the same court, and nearly opposite the window of my friend, my window also looked—I had therefore ample opportunity of observing the activity of this affection. To most men · of contemplative habits, the rude and noisy mirth of those ill-regulated, ill-clothed creatures would have proved distracting — not so to Brunel. To him it brought no disturbance, except when a cry of distress was heard. Then pen and pencil were abandoned, and the venerable head and active body of Brunel might be seen rushing to the rescue. Not satisfied with raising the little victim of petty tyranny from the gutter, he would sometimes bear it in his arms to his house, and never cease his caresses until its little heart was comforted, and its sorrows effaced. He was in the habit of carrying halfpence in his pocket for poor children. A nice-looking child would always win from him a kiss, as well as the halfpenny, " for the clean face." A dirty child would also receive the halfpenny, if it promised to go home and ask its mother to wash its face.

It may be in the recollection of some of the *savants* who attended the meeting of the British Association at Plymouth in 1841, to have witnessed this love for the young carried so far as not only to test their gravity, but almost to compromise their dignity. One day, upon entering the author's lodgings, there, like the great Edmund Burke, lay Brunel before them on the floor, in full romp of hide-and-seek with the author's children.

Lady Hawes writes : " With young people generally he was full of fun ; amusing them with a variety of sleight-of-hand tricks, in which he possessed considerable skill, and with stories also of an exciting character."

Amongst the emigrants who most constantly enjoyed her father's hospitality, about the year 1800, was Monsieur de `C., a gentleman of high family, agreeable manners, and considerable conversational powers. To these social recommendations was added that of being a hero of romance. " One remarkable passage in his early life was often recounted by my father to us," says Lady Hawes, " as the story of the *Green Room*." The incidents connected with this story were subsequently distorted into a melodrama for the English stage. But as the reality may be found to possess more interest than the fiction, I venture to reproduce the tale in its integrity, in the Appendix E.

To his own children he was a loving and devoted father, and much of the early success which attended the career of his distinguished son, must be attributed to the accuracy of the instruction and affectionate superintendence which that son received from his father during the early years of his life ; and of which we can offer no better proof than the fact, that when sent to school at eight years of age, young Isambard, was already familiar with the first books of Euclid, and had obtained some notions of the elements of mechanics. His mother, who had witnessed many of the uncertainties which are found to attend the profession of a civil engineer, sought to induce his father to select for him some other profession. At the time the question was under discussion, Dr. Spurzheim, the phrenologist, became a visitor in the family ; his opinion was naturally sought.   Upon an examination of the boy's develop- ment, he declared that it would be as useless to oppose the tendencies which they exhibited, as it was in his father's case.   This prediction appears to have been subsequently confirmed by M. Breguet, the celebrated chronometer and watchmaker of Paris, in whose *atelier* young Isambard gave the most unquestionable evidence of his mechanical aptitude, and from whom he obtained valuable instruction.   Writing to the father, November 1st, 1821, he says : " Je sens qu'il est important de cultiver chez lui les heureuses dispositions inventives qu'il doit à la nature, ou à l'éducation, mais qu'il serait bien dommage de voir perdre." Such proficiency, indeed, had he obtained in drawing, that " alphabet of the engineer," that when in the following year the designs by his father for a bridge to span the Neva at St. Petersburg were placed in his hands, that he might supply a few figures as a scale, and so become identified with his father in the design, the drawing was soon returned—not with a few figures only, but actually peopled with illustrations of every manual operation. Indeed no opportunity was omitted by his father in pointing out how and where instruction was to be obtained.   So late as 1823, and upon the anniversary of his birth, his father writes : —

" J'espère, mon cher enfant, que tu emplois bien ton temps.   Si tu es à Bowling, c'est là qu'il faut étudier la fabrique du fer, depuis le minerai jusqu'à une barre. Combien de charbon, de chaux, de minerai, combien de coke, enfin, chaque procédé, avec toute la précision d'un chimiste.   On ne te refusera rien là, et quand tu seras au fait tu pourras parler avec les fabriquants que tu verras par la suite. Aujourd'hui 17 ans ! Quel homme! Aux âmes bien nées la valeur n'attend point le nombre des années. Oui, mon cher Isambard, tu as une carrière ouverte, il faut en tirer un bon parti.

" Your dear nurse brought a very fine geranium : dear creature ! She thinks of no one but us and yourself."

Brunel's instructions were not, however, confined to the engineer's office or the father's study, nor to one member of his family only.

" Some of my most pleasing early recollections," says Lady Hawes, " are connected with country walks made with my dear father." In his fresh, warm admiration of nature, he more than sympathised with his children, while the knowledge which he was always ready to impart, by giving direction to their thoughts, added intensity to their wonder, and lead them ever more and more to lift their minds to the " wisdom and spirit of the universe."

Nor was the instinctive kindness of his nature restricted to the young.

It was during a visit to Ireland in 1833, and when travelling by coach from Cork to Waterford, that he took advantage of the change of horses at Youghal to stroll through the town. Observing a number of poor people in a state of distraction around something in the middle of an open space, he asked the cause.

" Sure, then, where will poor paple git the wather at all, at all, barring the salt wather from the say ? " was the reply. Some mischievous boys had injured the public pump ; no water was to be had ; nor could any one be found either competent or willing to remedy the evil. Here was sufficient to excite the active sympathies of

Brunel. Impressed only by the misery before him, to which he believed he could supply the remedy, he did not hesitate. His coat was instantly off. The pin of the handle of the pump was at once knocked out, the piston raised, and the choked valve rectified. In his delight at witnessing the happiness which he had thus spread around him, he utterly forgot the coach and the object of his journey. He cheerfully accepted his position, however, and consoled himself in being afforded the opportunity of forming an acquaintance with an Irish car, in which he was obliged to continue his journey.

The sufferings to which mail and stage-coach horses were subjected by the increase of speed exercised, he thought, a demoralising influence. He often alluded to the inhuman sacrifices coach proprietors were compelled to make to meet the increasing demand of the public with unfeigned regret, and no one felt more strongly than he the benefits which railways were calculated to confer, not only upon the agricultural and commercial interests of the country, but upon the moral character of the people.

"See," he would say, " there is Mr. Horne with his *four hundred horses;* in three years his fast coaches and the mails kill all he supplies to them. In the stage coaches they only last six years. And there is Mr. Waterhouse, he must renew his stock of *three hundred horses* in four years,—it is shocking! By increasing the speed from eight miles an hour to ten, he loses a horse in every run of 200 miles. Certainly," he would add, " to Mr. McAdam the poor horses are much indebted. In Sussex, since Mr. Campbell improved the roads with but an indifferent material, the coachmasters say that they break only *one* horse's knees now, where they used to break *twenty.*"

"In all the round
Of being, is there aught in God's pure eye
So blessed, so sanctified, as those kind thoughts
That stir the bosom of Benevolence ? "

The active interest which he exhibited in the proceedings of all societies having for their object the amelioration of the evils of life was well known. Respect for the dead was also a characteristic of his mind, and he would often contrast the condition of our teeming churchyards and desecrated burial-grounds, with the reverential beauty of Père-la-Chaise at Paris, its historical recollections, and its simple, tender and solemn inscriptions.

With Mr. Augustus Pugin, who supplied a variety of beautiful designs for mortuary chapels, gateways and monuments, he, in 1825, endeavoured to organise a company for the establishment of a large cemetery, to be called the NECROPOLIS OF LONDON ; but the idea of separating the place of burial from the place of worship was so much opposed to the habits and the prejudices of the people, that the project failed at that time to enlist the public sympathy ; nor until the evils and the horrors of intramural burial had been so ably exposed by Dr. Walker and others, were people's minds weaned from their time-honoured folly. Under an awakening sense of a sanitary necessity, a company was, some years after, formed, and KENSAL GREEN now imperfectly fulfils, in its heterogeneous collection of monuments, the beautiful, symmetrical and classical project of Brunel and Pugin.

The joint action of benevolence and love of approbation, which rendered Brunel courteous to all who approached him, subjected him often to severe trials of patience and forbearance. Inventions of the most immature character were constantly presented to him for examination and opinion.

Lady Hawes says: "I remember an Irishman once submitting to my father a drawing of a sort of hood for a carriage, which in fine weather was to hang *under* the body, ready for use. My father pointed out the impracticability of stowing away so large a mass in so small a space. The answer was ready, 'Oh! then it must be left at home in fine weather!'"

At another time his attention was requested to an invention for sweeping chimneys that should dispense altogether with climbing boys. A broom was to be worked from *above* as well as from below. To the natural question, "How was the rope to be got to the top?" the answer was, "Of course a little boy must go up with it first."

To a pressing invitation from a M. le Comte de Bec-de-Lièvre, to join him in a project which he declared was to produce extraordinary results, Brunel naturally required to know something of the process by which the Count's results were to be obtained, and also the nature of the results themselves. M. le Comte declining to satisfy him, he closed the correspondence thus: "Permettez-moi de vous dire franchement, que je ne puis m'occuper de votre proposition avant d'avoir des résultats, et avant de connaître le procédé;" and he slily adds, "j'ai eu occasion de voir bien des inventions ou découvertes tomber, malgré la confiance des auteurs."

"Ah! my friend," he would often say, "it is very easy to invent a machine, but it is not very easy to make it work."

However much these extreme cases of the vanity of inventors may have afforded subjects for social amusement, yet there were many which, though less absurd, did not prove more practical, pressed upon his time, and were felt to be great and unjustifiable intrusions.

But it was not enough that his courtesy should be thus abused. Schemes were placed before the public falsely announced as being sanctioned by Brunel's opinion, and invested with Brunel's authority.

A Mr. Collier, in 1815, invented a new machine, which he denominated the criopyrite, or fire-ram. The public attention was enlisted, and a Dr. Thompson wrote to Mr. Ellicombe, then Brunel's resident engineer at Chatham, for particulars. The letter was forwarded to Brunel. He was indignant, and in reply to Mr. Ellicombe says: " You would oblige me by answering the application of Dr. Thompson, by stating, as coming from me, that nothing is more preposterous than the account which has been published respecting this engine which, it is added, consumes no more than *one-twentieth* part of the fuel required for a steam-engine of the like power.

" It is true that an attempt has been made with a view of obtaining all those advantages which the author of that engine anticipated as certain. Having been called upon to witness its action, and to give my aid in directing its power, I am able to state that the new engine, supposed to possess a power equal to *twenty horses*, has not yet, to my knowledge, moved without some external aid of two or three men. The account given out is, therefore, a gross imposition, which you will, I hope, correct under my sanction."

Of works of real merit, Brunel was always a willing and judicious admirer. His perfect knowledge of me-

chanical construction enabled him at once to appreciate
the value of the objects sought to be obtained, and to
estimate the difficulties which were overcome ; his
opinion was, therefore, eagerly sought for, not only by
the mere schemer, but by the real inventor and im-
prover. "The pleasure he derived from the contemplation of a well considered piece of mechanism," says
Lady Hawes, "was intense, and his admiration honest
and fervent; never distorted by professional jealousy."
Of Mr. Heathcote's beautiful machine, which was to
supersede the laborious and complicated operations of
the lacemaker, as the stocking-loom had done that of
the knitter, he used to speak with unqualified admiration ; and he did not hesitate to declare, when called
upon to give evidence as to the validity of the patent,
that it appeared to him " one of the most complete
mechanical combinations," exhibiting "originality and
ingenuity," and displaying " uncommon powers of invention for the purpose of accomplishing a texture
which had been attempted before, but to his knowledge
without success ; " while to Mr. Heathcote, he predicted that he should live to see the day when lace,
then sold at *six guineas* a square yard, would become
as cheap as calico. The first square yard of plain net
was sold from the machine for *five pounds ;* but for the
last twenty-five years, the average price of a somewhat
inferior quality has been *five pence!* thus fully vindicating Brunel's anticipations.

At a time when the labours of Mr. Babbage were
little appreciated, or indeed scarcely understood, Brunel
did not hesitate to express his admiration for the
originality of conception, and the amount of mechanical
invention exhibited by his friend in his machine, or
difference engine, by means of which tabular numbers

were not only to be computed, but the results actually impressed upon metal plates, from which they might be directly printed on paper without resorting to the use of ordinary types. He saw the immense service which Babbage was about to confer upon practical science, upon the astronomer and the navigator, the engineer and the actuary, by freeing numerical investigations from the accumulated—overwhelming labour which tabulated error involved, and from which no published table was exempt.

He could appreciate in their entirety the scientific attainments, the mechanical skill, and the indomitable perseverance which such a work demanded. Indeed to the success of few inventions did Brunel look forward with more anxious solicitude than he did to that of Mr. Babbage. Although the intercourse between these eminent men was almost altogether personal, yet the few communications which have been preserved, show how strongly Brunel's sympathies were enlisted.

"When will you call upon me," writes Brunel, July 1st, 1829, " or tell us *in any way* how you are going on. We feel for the success of your distinguished labours; we, therefore, wish to know whether we can forward your views." And when Mr. Babbage was negotiating with the Government for a locality wherein to erect his machinery, there was no one more actively zealous in his service than Brunel. "Your stoppage on the other day," he says, " did not lessen the *perfect confidence* I have in the machine. I shall be glad to have a little to say as to the future proceedings before I see Lord Althorp."

That the sympathy which existed between these gifted men was great, I am assured by the solicitude evinced by Mr. Babbage for the restoration to him of the treasured documents from which I have been permitted to make the above extracts.

Nor was Brunel's interest confined to the application of mechanics to scientific questions, the arts also claimed his attention. Erard and other musical instrument makers not unfrequently benefited by his musical ear as well as mechanical skill, and friends on the Continent and in America applied habitually to him to select harps and pianos for them.

Although he was never able to overcome the difficulties of the English accent, he succeeded in writing the language with a great degree of facility and correctness. He could use it also in reply with perfect readiness. Lady Hawes mentions that when introduced to the Prince Regent (George IV.) during a visit to Woolwich, where Brunel had erected important works, the Prince observed, " *I* remember Mr. Brunel perfectly, but Mr. Brunel has forgotten *me*." " My father bowed respectfully, and expressed his regret that he should have been guilty of any omission or neglect. ' Yes,' continued the Prince, ' some years ago when you explained to me the wonders of the block machinery at Portsmouth, you promised me a copying machine of your invention, but you forgot your promise, Mr. Brunel.' Without hesitation or loss of presence of mind, my father rejoined, ' Please your Royal Highness, I have unfortunately never been able to perfect the machine, so as to make it worthy of your Royal Highness's acceptance.' "

Meeting one day an acquaintance of former years, he was saluted with pleased recognition, and with the exclamation, " Why, Brunel, how old you look ! but I suppose it must be ten years since we met." " Yes," said Brunel, " I have no doubt that it has been so long a time, if I may judge by the change which has been made in you." Lady Hawes farther relates, that once when her father was called upon to give evidence in a court of justice relative to a patent right, it was expected that some effort would be made to cast discredit upon his testimony, because he was not only a foreigner but a Frenchman ; and the question was asked in a super-cilious tone, in anticipation of a petty triumph : "You are a foreigner, Mr. Brunel ? " "Yes," replied Brunel, "I am a Norman, and Normandy is a country from whence your oldest nobility derive their titles." It is almost needless to add that this ready and spirited reply relieved Brunel from any farther impertinence.

The following answer to the Messrs. Borthwick of Leith, upon the feasibility of applying a small water-power to a variety of machinery : sawing wood, cutting stone and marble,· preparing oak bark, &c., is charac-terised by the same acuteness.

Those gentlemen had expressed to Brunel their ap-prehension that he would consider their notions extra-vagant, if not ridiculous, if they asked him to design machinery for such a variety of purposes, where the moving force was so limited. Brunel replied, " I see nothing ridiculous in your attempting to adapt a small power to many purposes, no more than to have many carts and only one horse. It is not expected you will fasten them all together when you want only the use of one."

" Of works of art," says Lady Hawes, " whether in painting or in sculpture, my father was both an enthu-siastic admirer, and an admirable critic. He would point out to us the style, the beauties, and the relative merits of different artists."

The few miniatures he has left of his own painting,

bear ample testimony to his talent in that department of art. The truthfulness of expression, and the beauty and finish of manipulation which they exhibit, drew from one of our first miniature painters exclamations of astonishment, and the opinion that had Brunel pursued the art as a profession, he would have been one of the most distinguished miniature painters of the day.

To his mind lines accurately represented forces, and of their relative value and position in a structure, he would always satisfy himself before accepting any numerical calculations. Hence the immense importance he attached to correct drawing : always considering it the true " *alphabet of the engineer*," without a knowledge of which he believed no complete idea of construction could be realised. The facility which he had himself obtained in expressing his ideas by lines, he retained to a very late period of his life ; and he often exhibited the accuracy of his delineation, by describing a circle with his hand only, and afterwards determining the centre with mathematical precision. He was unwilling to admit that this facility of manipulation was altogether to be attributed to natural development. Constant persevering practice he deemed essential to attainment in every art. For mere inspiration of genius he entertained but little respect. The constant striving after excellence which distinguished his own mind, never permitted him to rest satisfied with imperfect execution.

Imitating nature in her productions, he seemed always willing to bestow his labour upon the smallest accessory as upon the most elaborate conception, upon the threads of a screw as upon the movements of the most complicated machine, holding strongly to the opinion of his relative, N. Poussin, that "whatever was worth doing was worth doing well;" and any improvement, however small, that occurred to his mind, was at once submitted to the test of projection, whether by day or during the night. It therefore seldom occurred that he permitted himself to enjoy more than five hours' rest.

> " The height by great men reached, and kept,
> Is not attained by sudden flight ;
> But they, while their companions slept,
> Were toiling upwards in the night."

Lady Hawes mentions that one night this intellectual restlessness of her father saved the house from being plundered. He had risen to fix his conceptions, and while still occupied in his study, some burglars, who had entered the house to rob him of the products of his industry, became alarmed, left the greater part of their booty, and fled.

It has been made matter of censure that Brunel never adhered to an original estimate. The charge was urged at an early period by the Government, and more or less echoed by individuals ever after ; but this charge can scarcely be considered just. In many instances, those who consulted Brunel had such limited conceptions of the nature of their own requirements, that they were led to anticipate a corresponding limit in the cost of the work which they sought to have performed ; but where, with Brunel, excellence was the object, his suggestive and comprehensive mind induced an expansion of ideas in his employers, and, as a consequence, a desire to realise results which they could have never contemplated. These enlarged views demanded farther thought and more elaborate designs, but going so far beyond the original notions, they left an impression of Brunel's extravagance : where, however, the real object was to secure completeness, then were the suggestions of Brunel accepted in all their integrity without disappointment or regret.

Connected with this phase of character is the accusation of want of candour, a certain reticence in conveying his opinions, which created doubt as to his ultimate intentions, and which was by ordinary minds interpreted to be a want of truth.

That this unwillingness to come to a decision was characteristic of Brunel I entirely admit ; but that the cause was to be attributed to want of integrity of purpose I entirely deny. A better philosophy has taught me that the very faculty which in its exercise would prevent a hasty decision, forms really an important element in the constituents of genius as well as of wisdom.

To restrain the influence of first impressions, to subdue any outward expression of emotion until observation had secured all the necessary facts — the understanding had drawn the necessary inferences — and the constructive faculty had supplied the requisite means of accomplishing a given end, is the office of this faculty, which, not understood, assumes the form of want of candour. A great authority long ago said, " A fool uttereth all his mind, but a wise man keepeth it till afterwards."

The rapid judgment which Brunel's appreciation of the value of lines enabled him to pass upon the merits of any project, where he had the opportunity of examining the drawings, was most remarkable.

I remember well the morning he received a longitudinal section of the first chain bridge thrown across the Seine at Paris by M. Navier, the premier Ingénieur des Ponts et Chaussées of France, one of the earliest and best writers on suspension bridges, and a man distinguished for his physico-mathematical researches.

" Look here," exclaimed Brunel, as he examined the drawing.

" You *woold* not venture I think on that bridge unless you *woold* wish to have a dive ? "

" No," he added, " that will not stand, that will tumble into the river."

I observed that M. Navier had a high reputation for his mathematical knowledge and facility in arithmetical computations.

" Ah, well ! " replied Brunel, " may be ; but this time he has left out the last nought in his calculations."

Not long after we received an account of the fall of the bridge, said at first to have been caused by the bursting of a water-pipe, which softened the adjacent ground; but afterwards acknowledged to have been the result of faulty construction.

This anticipation, derived from his thorough confidence in geometrical projection, reminds me of a similar instance which occurred at Deptford, where a large store had been just completed, which we had to pass in our walk. As we approached the building, Brunel hastened his steps, saying, " Come along, come along ; don't you see, don't you see ? " To my interrogation as to the cause of his alarm, his reply was, " There ! don't you see ? It will fall ! " What was about to happen was as palpable to his mind as if it had happened. He had observed the want of perpendicularity in the structure, and the conviction was as strong upon him that it could not stand, as if he had seen it fall.   The next morning we learnt that the building was a ruin.

In 1826 the application of cast iron was receiving an unusual attention from engineers.   Mr. Maudslay determined to exhibit his confidence in the material by erecting a roof over his factory in Lambeth.   But so strong was Brunel impressed with the insecurity of any structure of the kind unless well combined with ties of wrought iron, that when a rumour reached him on the morning of the 24th May, 1826, that a serious accident had occurred at Maudslay's, he at once exclaimed, "The roof! the roof has fallen!"   And so it proved.

His capacity of applying knowledge once obtained was very remarkable.   The following anecdote, supplied by Lady Hawes, offers a striking illustration. Of the events recorded she was herself an anxious witness, and on her young mind they naturally made an indelible impression.   Early in the year 1817 Brunel visited Paris, upon the invitation of an English company, taking his family with him, in anticipation of a protracted sojourn.   The object of this visit was to suggest the best means of supplying the city with water, and being personally known to the King, to carry on the necessary negotiations with the Government.   Suffice it here to say, that much valuable time was expended to no purpose, and that Brunel found it necessary to hasten his return home.   When, in the month of December, he arrived at Calais, a violent storm prevented any vessel from quitting the port, and for five weary days he was compelled to control his impatience as he could.   On the sixth day he took advantage of some abatement in the weather to embark with his family on board the first vessel which offered to sail.   This proved to be French, with very

indifferent and totally insufficient accommodation. The
vessel had not proceeded far to sea when the weather
became again threatening, and as she approached the
English coast, so bad, that serious alarm began to be
entertained. The captain gave his orders with hesitation, confidence was shaken, and discipline endangered.
The proximity of Dover prompted him to run for that
harbour, contrary to the experience of his best men,
and to the strongly reiterated opinion of Brunel, that if
he ventured so rash an attempt nothing could save the
vessel from destruction ; still, in spite of all remonstrance and a heavy south-west gale, he held on his
course.

Brunel had not for twenty-four years exercised in
any way his early profession ; he had not, however,
forgotten the lessons then received ; and during this
fearful contest with the elements he had exhibited so
much nautical knowledge, as well as physical courage,
that all on board, sailors as well as passengers, prayed
that he would take command.

The obstinate, ignorant, and perverse captain resisted, until his cupidity was stimulated by a bribe,
when Brunel was permitted to take the helm. " To
Deal," he cried. The course was set, confidence was
restored, his instructions were obeyed with cheerfulness and alacrity, and the vessel, with its precious
cargo, brought in safety to port. Some of the vessels
which had also sailed that morning from Calais, having
no Brunel to direct the course, were beating about the
channel all night, while others were two days before
they reached their destination.

From the great difficulty which was experienced in
landing the passengers at Deal, it was obvious to those
acquainted with the coast, what amount of danger

would have been incurred had the attempt to reach Dover harbour been persisted in.

The grateful acknowledgment of passengers and sailors repaid the courage and the skill which had thus rescued them in the hour of danger.

Brunel's presence of mind and promptitude of action were early conspicuous. During his sojourn in America these valuable properties were often called forth. Once, for example, when employed on an island in Lake Champlain, he chanced to arouse the vindictive instincts of a rattlesnake. His companions fled, but Brunel stood his ground, and as the reptile approached he broke its back with a heavy stone skilfully thrown.

At a later period of his life, while in the act of inspecting the Birmingham railway, a train, to the horror of the bystanders, was observed to approach from either end of the line with a velocity which in the early experience of locomotives, Brunel was unable to appreciate. Without attempting to cross the road, he at once buttoned his coat, brought the skirts close round him, and firmly placing himself between the two lines of rail, waited with confidence the issue. The trains swept past leaving Brunel unscathed.

Impressed as Brunel's mind was with the superiority of modern civilisation, it was natural to suppose that he would be sometimes called upon to defend his opinions. When on a visit to the author in the neighbourhood of Cork in 1833, Mr. Joseph Leicester, the mayor, invited a select party to meet him at dinner. The conversation turned upon the comparative mechanical merits of the ancients and the moderns. A barrister of loud tone and pretentious manner, with whom the discussion originated, was the "laudator

temporis acti." After expatiating a considerable time
upon the wonderful productions of Grecian art and the
architectural glories of Imperial Rome; of the triumphs
of Phidias and Praxiteles; the boast of Augustus;
the magnificence of Vespasian; the genius of Trajan;
the public monuments of Hadrian; and the munifi-
cent patronage of the Antonines, Brunel quietly drew
from his pocket a chronometer-watch, and holding it up,
asked, " whether such a thing had ever been found at
Athens, Herculaneum, or Pompeii." The question was
so suggestive of the benefits which mechanical skill
had conferred on navigation, of that union of commerce
with science, without which there can be no bond
of nations, that all felt the evidence to be conclusive,
against any amount of forensic declamation.

He then referred to one circumstance, as exhibiting
a very low state of early civilisation, namely, the
want of sympathy for the suffering poor. " There
were no hospitals in those days," he exclaimed with a
sigh. In fact the word hospital never once occurs in
Gibbon's copious index to his " Decline and Fall."

Although Brunel was not what is called a man of
business,—much owing to the reliance he was constantly led to place on others in money matters,—much
owing to the suggestive, rather than the conservative,
character of his own mind—evidence the gentle remonstrance from the Navy Office as to the irregularity
of neglecting to furnish vouchers with his accounts;
and directing that accounts of tradesmen should be
opened directly with the Board, which undertook to
settle them when approved by Brunel;—yet his power
of calculation was unusually rapid; and to any work
emanating from a manufacturer's hands he could almost
at a glance assign its true value.

" I remember a lady," writes Lady Hawes, " once exhibiting to my mother a costly foreign lace, which she had just brought from the Continent, and which she prided herself upon having obtained a great bargain. My father examined it, and observing that the groundwork was formed of bobbin-net, executed by machinery, with which, as we have seen, he was already well acquainted, he at once gave to it its real value, to the inexpressible mortification of the lady, who found on explanation, that she had secured no bargain."

From one eccentricity, often found to characterise genius, Brunel was not free. The necessity of concentrating certain faculties of the mind upon one special object of thought, necessarily tends, for the time, so to weaken the influence of other faculties, that contact with the external world would seem to be retained only by a thread, and causing even atmospheric influences to be disregarded. This absorbing influence of thought in Sir Isaac Newton, would arrest him while about to rise, and retain him seated undressed on his bed for hours, in that profound contemplation which, according to his own admission, alone enabled him to wait the evolution of thought, and from the first dawn, " little by little, to attain to the full and clear light."

Although excellence can be scarcely looked for without the power of abstraction ; yet it offers, more than any other tendency of the mind, examples of perversion to the ludicrous. Under its influence, it is related that Newton was tempted to use a lady's finger as a tobacco-stopper ; Dr. Robert Hamilton, to take off his hat to his wife in the streets, and apologise for neglecting her salutation, as he had not the pleasure of her acquaintance ; the Rev. George Harvest, to go out gudgeon fishing when he should have appeared at the hymeneal altar with his bishop's daughter ; and Brunel to caress the hand of a lady to whom he was scarcely known, but who happened to be seated next him at table, believing it to be that of his own wife. In these fits of abstraction he would also mistake the day and hour of an invitation ; would go to the wrong house ; and, as once occurred at Lord Spencer's, actually subjected himself to be turned away from that hospitable door, because he sent up another person's card instead of his own.

Many times has he been known to take the wrong coach, and to find himself set down far in the country, when he ought to have been in town.

Leaving his umbrella behind was an almost invariable occurrence. " On one occasion," says Lady Hawes, " having been specially reminded by my mother not to forget it, he kept it in his hand during the whole time of a visit to a friend. Upon taking leave, his eye fell upon another umbrella in the room strongly resembling his ; of that he took possession, and with it protected himself during his walk ; nor until he reached home was he once conscious that his own was under his arm."

Brunel was, as I have said, often compelled to place his pecuniary interests in the hands of those whose want of capacity, or equivocal integrity, more than once brought him to the verge of ruin, from which he was only saved by an indomitable energy of mind, equanimity of temper, and incompressible elasticity of spirit, that no opposition could interrupt, and no misfortune subdue.

Mr. Josiah Field mentioned to me that when the great saw-mills at Battersea were destroyed by fire, he was with Brunel at Chatham when a messenger arrived to announce the catastrophe, but who evidently dreaded to make the communication. Brunel, however, asked only one question, " Is anybody hurt ? " The answer being in the negative, he turned to Mr. Field, and said, " I can make better machinery now." Like Audubon, the American ornithologist, who, though at first entirely overcome upon witnessing the havock which two Norway rats had made of two hundred of his original drawings, the labour of years, did not give way to despondence, but with hopeful cheerfulness took up his gun, his note-book, and his pencils, and went forth to the woods, feeling pleased that he " might now make better drawings than before." So Brunel, in the same spirit, could write in reply to a letter of condolence from Mr. Edgeworth, upon his losses at Battersea. " The misfortune is not without its consolation, as I shall now have the opportunity of carrying out many improvements which I have often contemplated."

And when congratulated upon having successfully overcome all those difficulties which beset him at the Tunnel, he would reply, " Néanmoins, si je l'avais à refaire je ferais mieux."

Finally, it will, I think, be admitted that, notwithstanding all his weaknesses and peculiarities, Brunel possessed powerful elements of character, and none more striking than that constancy of purpose, and that steady confidence in his own resources, which never permitted him to doubt of ultimate success, often in despite of cruel disappointments and the heavy pressure of apparent defeat.

In the summary which M. Edouard Frère has given of Brunel's life, he says : " Sir Isambard Brunel a mérité par l'éclat de ses inventions ; par la dignité d'une carrière vouée tout entière au travail ; par l'élévation de ses

vertus privées, la célébrité qui entoure son nom, l'admiration de tous les hommes de savoir et de labeur, et
le souvenir affectueux de tous ceux qui, assez heureux
pour le connaître personnellement, ont pu apprécier son
caractère à la fois si simple et si noble." *

To this just and discriminating résumé, I shall only
add that the unaffected piety by which the close of his
laborious life was cheered as well as soothed left
nothing more to be desired : a life honourable in
itself, and rich in the practical benefits which it conferred upon the country of his adoption.

* Notice historique sur la Vie et les Travaux de Marc Isambard
Brunel, par Edouard Frère.

CHAPTER XX.

CONCLUSION.

1842–1849.

PROFESSIONAL CAREER TERMINATED, 1842 — STACKING TIMBER IN
DOCKYARDS — SUCCESSES OF HIS SON — HER MAJESTY'S VISIT TO
THE TUNNEL, 1843 — SOCIETY OF HIS GRANDCHILDREN — SECOND
ATTACK OF PARALYSIS, 1845 — EQUANIMITY AND CHEERFULNESS
UNDER PHYSICAL SUFFERING — DEATH, 12TH DECEMBER, 1849.

WITH the completion of the tunnel Brunel's professional career must be considered to have terminated; for with the exception of a plan for STACKING
TIMBER IN DOCKYARDS, which had been suggested by him
in 1824, and laid aside for the more important project of
the tunnel, he steadily declined all offers of employment
connected with any of the great undertakings of the day.
As the drawings which he submitted to the Admiralty,
explanatory and illustrative of his plan, were, I understand, never returned, and as the beauty and economy
of the arrangement were recognised by some in authority, and more particularly by Admiral Sir Hyde Parker,
as well as by the master shipwright at Portsmouth, it
is possible that the attention of the Admiralty may yet
be called to the subject. Sir Hyde Parker, in expressing
his regret at " there not being a probability of its being
carried into effect," though " quite certain it would
have suited this dockyard (Portsmouth) admirably,"

adds, "unfortunately prejudices take place without reason." Thus to the last, opposition continued to present itself to Brunel's progressive spirit; but that spirit had learnt to look with tranquillity upon disappointment.

To become personally acquainted with some of those successes which had been achieved by his son, was now to Brunel his most gratifying relaxation. On the 19th of July, 1843, he had the pleasure to witness the launch of the Great Britain steamer, of 3500 tons, at Bristol; in which, for the first time, the screw as a propeller was applied to a vessel of large burden.

In his journal, Brunel, while noting his sympathy with this great progressive movement in naval architecture, adds : " Many years ago I made trials of various means for propelling boats, and in order to ascertain the degrees of effect, I had a circular canal prepared, in which the various models could be made to operate with great precision : *the screw was one of the means of propulsion.*"

During this excursion he was tempted to prolong his absence from London. He thereby failed to receive the announcement that her Majesty had expressed her gracious intention to honour the Thames Tunnel with a visit.

On the 26th, of July, 1843, and while Brunel was in Somersetshire, the Royal visit was made. It was felt as a grievous disappointment to Brunel, that he had been deprived of the honour of receiving her Majesty, and of affording those explanations which would naturally have been required. And from the reference which he more than once makes in his journal to that eventful visit, it is evident that few rewards would have been so highly prized by Brunel, as that of receiving from the lips of his sovereign the expression of her gracious approbation.

Returned to London, he took up his residence in a small but cheerful house in Park Street, Westminster, fronting St. James's Park. Here he was cheered by the pious generosity of his son, and supported by the watchful and affectionate care of his devoted wife

He had lived to witness the happiness and prosperity of his children, and he was now gladdened by the companionship of his grandchildren, who became his constant companions. In his walks he would point out to them the wonders of nature, as he had done in earlier days to his own children, and feeling, it may be with deeper reverence, that

> " There is to him who reads the sacred page
> With knowledge, faith, humility, and love,
> A sweet and balmy influence in creation,
>
> *   *   *   *   *
>
> Winning the weary heart from earthly ills,
> Filling the mind with proofs of love divine,
> And pointing onwards to eternity."

In 1845 he was deprived, by a second attack of paralysis, of that society which he so much loved — the society of children. Still his equanimity and thoughtfulness for others never forsook him. In him was, indeed, "that constant flow of love that knew no fall," cheerfully adopting every suggestion for fixing a pen in his powerless hand ; and only smiling at the failures, he resolutely set himself to work to render his left hand available, in which effort he to a considerable extent succeeded. "So free was he from peevishness or fretfulness, and so thankful was he for the smallest attention, that the nurse who attended him," says Lady Hawes, "often declared it a pleasure to serve him."

z

He now withdrew altogether from the excitement of general society, perfectly contented with that of his own domestic circle. With the ruling passion strong to the end, he would strive to give expression to the thoughts ever working within him. The spirit was still there in undiminished integrity, but the power to realise its conceptions was no more. Unconscious of all around him, his hands would move as though they were embodying his thoughts. Gradually his hold on the external world relaxed, and though utterance failed, it was not difficult to comprehend the elevation of " the spirit's noiseless prayer."

On the 12th of December, 1849, and in the 81st year of his age, at peace with himself and all beside, he calmly sank to rest, leaving a name to be cherished so long as mechanical science shall be honoured.

On the 17th of December his remains were deposited in Kensal Green Cemetery.

In thus tracing the career of this distinguished and amiable man, I have not sought to elevate him to a higher position in the temple of fame than I believe his character and his works justified ; and though it may appear anomalous to speak indifferently of a man's power, and of his weakness, yet, if a true picture is to be drawn of human nature, we must be satisfied to forego ideal harmony and beauty in the portrait, for the more valuable element of truth in the delineation. In the case of the subject of these memoirs, to have cast any of his distinctive qualities into the shade, would have been to present some other character to the world than that of MARC ISAMBARD BRUNEL.

APPENDICES

APPENDIX A.

" Be it remembered, that at a stated District Court of the United States, held for the District of New York, at the City of New York, in the said District of New York, on Tuesday, the second day of August, in the year of our Lord one thousand seven hundred and ninety-six, Marc Isambard Brunel came into the said Court and applied to the said Court to be admitted to become a Citizen of the United States of America, pursuant to the directions of the Act of the Congress of the said United States, entitled: 'An Act to establish an uniform rule of naturalisation, and to repeal the Act heretofore passed on that subject;' and the said Marc Isambard Brunel having thereupon produced to the said Court such evidence, and made such declaration and renunciation as by the said Act is required, it was considered by the said Court that the said Marc Isambard Brunel be admitted, and he was accordingly admitted by the said Court to be a Citizen of the United States of America.

" In testimony whereof the seal of the said Court is hereunto affixed.

" Witness, John Laurence, Esquire, Judge of the said District, at the City of New York, in the said District of New York, this second day of August, in the year of our Lord one thousand seven hundred and ninety-six, and of the Independence of the said United States, the twenty-first."

APPENDIX B.

———+———

DESCRIPTION OF THE BLOCK MACHINERY.

THE block machinery is divided into four classes.

1st. Sawing machines, large and small, the former applied to convert elm timber into proper dimensions to be submitted to other specific machines; and the latter to cut the lignum vitæ for the sheaves.

2nd. Machines for the manufacture of the shell of the blocks.

3rd. Machines for forming the *sheaves*.

4th. Machines for forming the iron pins for the blocks.

To these are added a large machine for boring parts of the very large blocks, called *made* blocks, and which cannot be wholly executed by machinery, and two machines for turning *dead-eyes*, or blocks without sheaves, used to attach the ship's shrouds to her sides.

The sawing machines are of two kinds, *reciprocating* and *circular*. The former may be viewed as a gigantic representative of the carpenter's hand-saw; the latter, with the circular knife, is the subject of a special patent, and is an invention by means of which valuable woods, as mahogany, rosewood, &c. &c., are economised in an unlooked-for manner. By the substitution of *veneers*, or slips of those heavy woods glued on deal, for the cumbersome solid masses formerly used, the beauty of surface is secured, while the product is multiplied *fifty-fold*.

These saws may be made to turn either in horizontal, vertical, or inclined planes, and are so arranged that all the pieces cut from the same log, may be produced precisely of the same thickness; while the working of the machines, far from demanding the attention of the skilled workman, can be consigned to that of an ordinary labourer.

Next in order to the saw-cutting machine, comes the boring machine. Three of these acting simultaneously perforate the scantling or piece of timber prepared by the saw—and form the holes which are to contain the centre pin for the sheaves of the block, and which become the commencement of the several mortises to contain the sheaves. The blocks are now passed to the mortising machine, where the holes made by the boring machine are *elongated* to the required dimensions.

Three *circular* saws now cut the angles of the blocks, and three shaping engines form *each* the outside surface of ten blocks at the same time.

Two scoring engines next form the grooves round the blocks to receive the ropes or straps by which they are suspended.

The blocks are smoothed and polished by hand.

For making the sheaves there are fourteen machines. The tree of lignum vitæ is first cut into pieces of the required thickness by what are termed *converting machines,* of which there are three, one a reciprocating saw, the two others circular saws.

Three pieces are made circular, and the centres pierced by two *rounding* and centring machines or *trepan saws.* A hole is then formed in the centre of each sheave to receive the coak or piece of brass or bell-metal to form the socket for the centre pin of the block.

Two riveting hammers secure the coak in place.

In some kinds of sheaves three holes are drilled through sheave and coak by a drilling machine, to receive short wire pins cut by the cutting shears, which are riveted down by riveting hammers. The centre holes through the coaks are now broached out to a true cylinder by three *broaching engines.*

And lastly, the faces and edges of the sheaves are turned to a flat surface in three *facing* lathes, which also form the grooves round their eyes for the rope.

The iron pins are forged by two smiths in the usual manner; the heated iron being placed between two tools called swages, each having a semi-cylindrical cavity, so that when hammered together they form a cylinder.   The *pin-turning lathe* renders the pin true, and the *polishing machine* completes the operation.

For forming the *dead-eyes* there are two machines.

Here we have forty-three machines constituting a system of machinery, each part executing its purpose with a precision, rapidity, delicacy and power never before exhibited.

APPENDIX C.

——◆——

COPYING PRESS, 1820.

A WOODEN box contained the *damper*, or small metallic cylinder, around which a number of sheets of calico, fine linen, or other light material of the size of the paper used, was rolled. This cylinder was deposited in a tube, and kept moist.

The press consisted of a gun-metal bottom having two standards, one at either end, of the same metal. To these standards were attached two levers, each having its fulcrum in the opposite standard. These levers acted uniformly on the head of a wooden casing, with a convex top and flat bottom. To the centre of the top was attached, by a screw, a thin steel plate spring, curved so as to correspond with the convexity of the wooden casing, and which, therefore, retained the casing in an elevated position when not acted on by the levers; this casing being free to move up and down between the standards, while the ends of the spring rested in apertures cut in the standards. By this arrangement, when the case was pressed down by the levers, the spring being kept up at each end was forced into a horizontal position, but ready as it resumed its curved form to raise the casing the instant the pressure was removed.

The original writing being placed in a book, as many leaves of thin paper, to which it was to be transferred, as there were pages in the original writing were turned over it; then from the cylinder of the damper, the same number of sheets of calico were unrolled and laid over the blank paper, and, finally,

on each of these was placed a sheet of oil paper.   The book
was now put into the transferring box beneath the case, the
levers were brought to act upon the head of the screw, by
which the steel spring was attached to the convex top of the
case; the case was pressed down, the book was drawn through,
undergoing in its progress a succession of pressures, and thus
the writing was transferred.

APPENDIX D.

———✦———

IMPROVEMENTS IN MARINE STEAM ENGINES, 1822.

THE rotatory action was accomplished by placing the line of power of the engines at right angles with each other; both piston-rods being made to act alternately on a solid crank by which they were connected.

The regulation of the engine was effected by means of an hydraulic apparatus, consisting of a small pump connected with the condenser which worked a plug, ingeniously contrived to modify the direct action of the governor, and thus effectually to prevent all shock to the engine; while for the power of gravitation, by which the balls of the governor are usually closed, there was substituted centrifugal force, by means of a spring connecting the governor by a lever with the moderator.

The third improvement was in providing and securing a constant supply of fresh water, by the use of a condenser, which should so economise the generated steam as to retain the first supply of water with scarcely any loss.

This condenser consisted of a combination of pipes surrounded by cold water, and which collectively formed a spacious chamber. In the steam-pipe was placed a safety valve opening into the condenser, so that any excess of steam was at once secured; and as the waste-pipe also entered the condenser no steam was permitted to escape. The condensed water was returned in the usual manner to the boiler by a small pump. The pump of the moderator drew from the condenser, at every stroke, a certain quantity of the heated water, and allowed an equal portion of cold to enter.

The condenser was also supplied with an air-pump, and thus, while the condensation was carried on without waste, the vacuum was preserved. The boilers were cylindrical with spherical ends, and to prevent, as far as possible, the agitation of the water in them, when the vessel was at sea, they were always to be kept full. To the tops of the boilers "*steam rooms*" of sufficient capacity were adapted. By merely perforating the tops of the boilers with small holes, the communication with the steam rooms was accomplished.

The fourth improvement was in the manner of supplying coal to the fire. This was effected by means of revolving cast-iron cylinders, connected with hoppers. The cylinders were pierced with apertures, through which a certain quantity of ground coals was distributed uniformly over the fire.

APPENDIX E.

———+———

STORY OF THE GREEN ROOM.

MONSIEUR DE C—— related that, when a very young man, being on his return home to France, by the south of Italy, from a tour through the southern parts of Europe, he took the opportunity to pay a long promised visit to an uncle, an ecclesiastic. Although personally a stranger to his relative, his reception was full of kindness and cordiality, and so agreeably did the time pass, that from day to day he was induced to prolong his visit.

At length his departure could no longer be deferred. Travelling at that period was a matter of physical difficulty. There were no public conveyances; the roads were so bad as to be sometimes impassable; and the inns were execrable. The few travellers whom necessity, the love of knowledge, or of the marvellous tempted from home, performed their journeys on horseback, and on horseback, therefore, M. de C—— must travel.

When he laid the programme of his route before his relative, the good man shook his head. " You will scarcely reach your first day's halt before nightfall," said he. " The road is not safe; the forest through which it passes is lonely, and notorious for the boldness and cruelty of the bandits who infest it. To travel through it alone would be madness; in fact, I should consider myself responsible to your father if I gave my sanction to your arrangement."

The idea of an adventure was, however, rather agreeable

than otherwise to the young traveller—he assured his uncle that he had no fears; that if attacked, he was well able to defend himself, and having already prolonged his absence beyond the time fixed, his father would be greatly disappointed, and however much he must regret to quit his uncle's hospitable roof, a sense of duty urged him to delay no longer. The abbé at length unwillingly acceded to his nephew's wishes, but only on condition, that an old and trusty servant should accompany him. "True, it is long since Antonio cast off the uniform of a soldier," said the old gentleman, "but you may rely upon his courage and his prudence." To this condition there could be no objection, and accordingly, after having received the parting blessing of his uncle, the young man set out, well armed, and accompanied by the trusty Antonio. It was the spring of the year, and early flowers and budding leaves gave a charm to the landscape, and the delicious freshness of the air imparted elasticity and vigour to the travellers.

> " Sweet April, many a thought
> Is wedded unto thee, as hearts are wed,
> Nor shall they fail, till to its Autumn brought,
> Life's golden fruit is shed."

Antonio proved an amusing and instructive companion. He had seen much of life; had passed through many a hard fought field, and had borne part in many a hair-breadth 'scape.

Noon was some hours past, when, "still plodding through tangled forest, and through dangerous ways," the travellers found themselves, with horses tired, and their distance from the town, at which they were to make their first night's halt, still unascertained. For the first time it occurred to the youth that the good uncle's anticipation might possibly be realised. Mile after mile was passed with increasing difficulty. The tall pines seemed about to close the way, now reduced to a mere path. The only sound which at long intervals reached their ears, was that of the woodman's axe. To obtain refreshment for their horses, and shelter for themselves, became every moment matter of necessity.

Hitherto there had been no indication of abode of any kind,

and they had almost despaired of finding shelter, when a bluish smoke was observed rising, at no great distance, into the calm evening air, and offering to them a signal of relief. Turning off the direct track, they found themselves presently before what appeared to be an inn, though by no means of an inviting character.

Whatever their disappointment, there was no room for choice, and they accepted the invitation of "mine host" to enter, while their horses should be baited, with the intention of reaching their destination before night. The landlord, affecting "a saucy roughness," expressed his fear that, however glad he'd be to forward the noble cavalier on his journey, seeing that his house could not offer the accommodation his distinguished guest must require; yet, as his excellency seemed to be a stranger, and not well acquainted with the dangers of the road, or the distance to ——, he must forgive his bluntness if he ventured to say that some hours might be spent before they could clear the forest, where they would be exposed to dangers the most fearful, against which neither their arms nor their courage could protect them, and from which he had himself more than once with difficulty escaped.

These representations had their effect. M. de C—— took counsel of Antonio, who could only suggest that bad accommodation and miserable fare were, at all events, to be preferred to an unequal struggle, in a close bound forest, with an invisible but cruel and relentless enemy, who was seldom known to perpetrate the crime of robbery without adding to it that of murder. With ill-disguised reluctance the young man relinquished his intention to proceed; so, after bestowing their first care on their horses, master and man returned to the house, resolved to accept cheerfully what could not be avoided.

The only domestic who appeared was a girl, whose youthful, anxious countenance, naturally awakened a feeling of interest in the heart of the young traveller.

"Ninetta, you will show this noble gentleman to the *Green Room*," said her master. Up a rough and dirty staircase M. de C—— and Antonio followed their gentle guide, into a

room containing two comfortless looking bedsteads, the dingy
curtains of which fell like palls around the beds. These, with
a few chairs and an unplaned table, constituted the whole
furniture of the apartment. Here the girl left them. The
landlord was now heard to leave the house, and the girl returned to the room, where she busied herself in preparing the
beds; exhibiting, however, a restlessness of manner, and an
evident desire to attract the attention of the young cavalier
which could not be mistaken. M. de C—— approached her.
In a low whisper she warned him that the intention was to
rob and murder him and his servant, and in proof of her assertions she, in a hurried and trembling voice, told him that
the landlord would immediately return to the house, to say
that some travellers, who were in the habit of journeying that
way, had just arrived, and having no other room to offer
them, he would ask whether the cavalier would permit them
to sup with him. Should he consent those supposed travellers
would take the opportunity to send his servant out of the
room, and in his absence would overpower, rob, and murder
him: should he refuse they would effect their object during
the night. She farther informed him that the distance from the
town was only half a league, and which he might have easily
reached; but now it was too late, warning had been given to
the band, and they must protect themselves. Then trembling
at the thought of how she had endangered her own life, by di-
vulging the secrets of the prison-house, she prayed him not to
betray her. M. de C—— could only express his grateful thanks
— assuring her that nothing should induce him to compromise
her — and thus reassured, she left the room. What was to be
done? Even if they succeeded in effecting their escape from
the house, they would certainly be intercepted. The danger
must therefore be met where they were. Assuming the girl's
story to be true, Antonio recommended that they should take
the initiative, and not defer their defence to the night.

He suggested that the men should be admitted, and the
seats so placed at table, that the most powerful of the band
should be opposite to his young master, the two others on
either side, and opposite one another. That he should not

leave the room on any pretence whatsoever; and that after he had placed the cheese upon the table with the dessert, his master was to ask for a glass of water, when he would put his pistol into his hand, which he should at once discharge at the man opposite to him. "For the others," said Antonio, "leave them to me."

It may well be imagined that a youth of nineteen naturally recoiled from the idea of shedding blood under such equivocal circumstances. The girl's story might prove to be untrue. Would he not then himself be the guilty party? What would his father, his uncle, think, if he were to return home with the stain of murder on his name? As all these ideas crowded on his mind he hesitated to adopt Antonio's plan.

The landlord's foot was now heard on the stairs, and entering the room, he apologised for his intrusion by stating that three travellers, old customers, who made a point, when journeying that way, to pass a night at his unworthy house, had just arrived,—that he was very unwilling to send them away at such a time of night, unrefreshed,—that he lamented to say he had no other room with which to accommodate them, and prayed that his distinguished guest would permit the gentlemen to sup with him. Such condescension would render him ever grateful.

The poor girl's story was but too faithfully corroborated. M. de C—— could no longer doubt. The real nature of their position became painfully apparent. The landlord's request was granted. The supper was placed on the table, and the three travellers were introduced. So forbidding was their aspect, that even without the previous warning they must have inspired distrust. The chairs were occupied as Antonio had arranged. The meal proceeded, but could not boast of many social elements. One observed that the young cavalier did not eat; another discovered that he had left his snuff-box below, might not the servant fetch it? Various clumsy expedients were tried, but in vain, to rid themselves of the presence of the stern-looking old soldier, which only tended to confirm in its full extent the prophetic warning of the girl.

A A

The only reply which Antonio condescended to make was to knock on the floor, that the landlord might execute the wishes of his friends. No landlord, however, appeared; every action of the men tended to increase apprehension, and to satisfy the young man that he would be justified in following Antonio's advice. Yet, as the time for taking the first dreaded step approached, he hesitated. There was, as yet, no proof to him of these men's intentions, nor of their previous guilt, to relieve him from the charge of deliberate murder. It was only the resolute appearance of Antonio, and the obvious collusion of the men, whose furtive glances were from time to time cast upon him, which restored his self-possession and confirmed his resolution.

The supper was nearly finished. The cheese and dessert were placed upon the table. The time for deliberation had passed, that for action had arrived. The glass of water was called for. Antonio slid the cocked pistol into his young master's hand; — the trigger was pulled, and the most powerful of the three men fell; the ball had pierced his heart. Meanwhile Antonio, true to his promise, stepping close behind one of the other men, discharged his pistol at him who was opposite, whom he mortally wounded; and then throwing himself upon the third, quickly disarmed him, and, with the aid of his young master, secured his hands and feet with cords, which were found in the apartment, and which were evidently intended to be used against themselves. Fearing a rescue, the pistols were quickly reloaded; but all was still below. No landlord appeared to ascertain the cause of such violence, and the only sounds that broke the silence of the night were the groans of the wounded man, and the execrations of him who had been secured. To gain the neighbouring town, and to communicate with the authorities, was the next consideration. It was agreed, that while M. de C—— undertook to seek for aid, Antonio should remain to guard the prisoners. Cautiously descending the staircase, where danger might be still lurking, M. de C—— made his way to the stable. Neither landlord nor maid was to be seen. Not a sound disturbed the tranquil air, and being now assured of the direction and the distance,

M. de C—— galloped to the ·town, which he found, as the
girl had stated, to be but a short distance from the outskirts
of the forest. He gave information, and the authorities,
having only too much reason to believe the truth of his story,
at once directed gendarmes to accompany him to the inn.

The three men were recognised as notorious brigands. The
poor girl, who had hid herself in the loft over the stable,
finding the young traveller safe, came from her hiding-place,
willingly aided the investigations of the gendarmes, and was
able to point out where thirteen travellers·had been buried,
after having been robbed and murdered by the very wretches,
from whose hands she had been so instrumental in rescuing
M. de C——. The landlord was subsequently arrested, and
his complicity being proved, his house was razed to the
ground, and he, with the surviving bandit, hung.

APPENDIX F.

—✦—

LIST OF PATENTS.

A.D.
1799. Machine for Writing and Drawing.
1801. Ships' Blocks.
1802. Trimmings and Borders for Muslins, Lawns and Cambrics.
1805. Saws and Machinery for sawing Timber.
1806. Cutting Veneers.
1808. Circular Saws.
1810. Shoes and Boots.
1810. Obtaining Motive Power.
1812. Saw-mills.
1813. Saw-mills.
1814. Rendering Leather durable.
1816. Knitting Machine.
1818. Forming Tunnels, or Drifts, underground.
1818. Manufacture of Tinfoil.
1820. Copying Press.
1820. Stereotype Printing Plates.
1822. Marine Steam Engines.
1825. Gas Engines.

LONDON
PRINTED BY SPOTTISWOODE AND CO.
NEA-STREET SQUARE

INDEX